Studies in Computational Intelligence

Volume 654

Series editor

Janusz Kacprzyk, Polish Academy of Sciences, Warsaw, Poland
e-mail: kacprzyk@ibspan.waw.pl

About this Series

The series "Studies in Computational Intelligence" (SCI) publishes new developments and advances in the various areas of computational intelligence—quickly and with a high quality. The intent is to cover the theory, applications, and design methods of computational intelligence, as embedded in the fields of engineering, computer science, physics and life sciences, as well as the methodologies behind them. The series contains monographs, lecture notes and edited volumes in computational intelligence spanning the areas of neural networks, connectionist systems, genetic algorithms, evolutionary computation, artificial intelligence, cellular automata, self-organizing systems, soft computing, fuzzy systems, and hybrid intelligent systems. Of particular value to both the contributors and the readership are the short publication timeframe and the worldwide distribution, which enable both wide and rapid dissemination of research output.

More information about this series at http://www.springer.com/series/7092

Roger Lee
Editor

Software Engineering Research, Management and Applications

Editor
Roger Lee
Software Engineering and Information
Central Michigan University
Mount Pleasant
USA

ISSN 1860-949X ISSN 1860-9503 (electronic)
Studies in Computational Intelligence
ISBN 978-3-319-81628-9 ISBN 978-3-319-33903-0 (eBook)
DOI 10.1007/978-3-319-33903-0

Softcover reprint of the hardcover 1st edition 2016

Printed on acid-free paper

This Springer imprint is published by Springer Nature
The registered company is Springer International Publishing AG Switzerland

Foreword

The purpose of the 14th International Conference on Software Engineering, Artificial Intelligence Research, Management and Applications (SERA 2016) held on June 8–10, 2015 at Towson University, USA is bringing together scientists, engineers, computer users, and students to share their experiences and exchange new ideas and research results about all aspects (theory, applications, and tools) of software engineering research, management and applications, and to discuss the practical challenges encountered along the way and the solutions adopted to solve them. The conference organizers selected the best 13 papers from those papers accepted for presentation at the conference in order to publish them in this volume. The papers were chosen based on review scores submitted by members of the program committee and underwent further rigorous rounds of review.

In Chap. "Human Motion Analysis and Classification Using Radar Micro-Doppler Signatures", Amirshahram Hematian, Yinan Yang, Chao Lu, and Sepideh Yazdani present a novel nonparametric method to detect and calculate human gait speed while analyzing human micro motions based on radar micro-Doppler signatures to classify human motions.

In Chap. "Performance Evaluation of NETCONF Protocolin MANET Using Emulation", Weichao Gao, James Nguyen, Daniel Ku, Hanlin Zhang, and Wei Yu leverage the Common Open Research Emulator (CORE), a network emulation tool, to conduct the quantitative performance evaluation of NETCONF in an emulated MANET environment.

In Chap. "A Fuzzy Logic Utility Framework (FLUF) to Support Information Assurance", E. Allison Newcomb and Robert J. Hammell discuss the use of fuzzy logic for accelerating the transformation of network monitoring tool alerts to actionable knowledge, suggest process improvement that combines information assurance and cyber defender expertise for holistic computer network defense, and describe an experimental design for collecting empirical data to support the continued research in this area.

In Chap. "A Framework for Requirements Knowledge Acquisition Using UML and Conceptual Graphs", Bingyang Wei and Harry S. Delugach present a knowledge-based framework to drive the process of acquiring requirements for each UML model. This framework is based on a central knowledge representation, the conceptual graphs.

In Chap. "Identification Method of Fault Level Based on Deep Learning for Open Source Software", Yoshinobu Tamura, Satoshi Ashida, Mitsuho Matsumoto, and Shigeru Yamada propose a method of open-source software reliability assessment based on the deep learning. Also, they show several numerical examples of open-source software reliability assessment in the actual software projects.

In Chap. "Monitoring Target Through Satellite Images by Using Deep Convolutional Networks", Xudong Sui, Jinfang Zhang, Xiaohui Hu, and Lei Zhang propose a method for target monitoring based on deep convolutional neural networks (DCNN). The method is implemented by three procedures: (i) Label the target and generate the dataset, (ii) train a classifier, and (iii) monitor the target.

In Chap. "A Method for Extracting Lexicon for Sentiment Analysis Based on Morphological Sentence Patterns", Youngsub Han, Yanggon Kim, and Ikhyeon Jang propose an unsupervised system for building aspect expressions to minimize human-coding efforts. The proposed method uses morphological sentence patterns through an aspect expression pattern recognizer. It guarantees relatively higher accuracy.

In Chap. "A Research for Finding Relationship Between Mass Media and Social Media Based on Agenda Setting Theory", Jinhyuck Choi, Youngsub Han, and Yanggon Kim analyze important social issues using big data generated from social media. They tried to apply the relationship between agenda setting theory and social media because they have received social issues from official accounts like news using social media, and then users shared social issues to other users; so they choose tweets of Baltimore Riot to analyze.

In Chap. "On the Prevalence of Function Side Effects in General Purpose Open Source Software Systems", Saleh M. Alnaeli, Amanda Ali Taha, and Tyler Timm examine the prevalence and distribution of function side effects in general-purpose software systems is presented. The study is conducted on 19 open-source systems comprising over 9.8 million lines of code (MLOC).

In Chap. "Object Oriented Method to Implement the Hierarchical and Concurrent States in UML State Chart Diagrams", Sunitha E.V. and Philip Samuel present an implementation pattern for the state diagram which includes both hierarchical and concurrent states. The state transitions of parallel states are delegated to the composite state class.

In Chap. "A New and Fast Variant of the Strict Strong Coloring Based Graph Distribution Algorithm", Nousseiba Guidoum, Meriem Bensouyad, and Djamel-EddineSaidouni propose a fast algorithm for distributing state spaces on a network of workstations. Our solution is an improved version of SSCGDA algorithm (for Strict Strong Coloring based Graph Distribution algorithm) which introduced the coloring concept and dominance relation in graphs for finding the good distribution of given graphs.

In Chap. "High Level Petri Net Modelling and Analysis of Flexible Web Services Composition", Ahmed Kheldoun, Kamel Barkaoui, Malika Ioualalen, and

Djaouida Dahmani propose a model to deal with flexibility in complex Web services composition (WSC). In this context, they use a model based on high-level Petri nets called RECATNets, where control and data flows are easily supported.

In Chap. "PMRF: Parameterized Matching-Ranking Framework", Fatma Ezzahra Gmati, Nadia Yacoubi-Ayadi, Afef Bahri, Salem Chakhar, and Alessio Ishizaka introduce the matching and the ranking algorithms supported by the PMRF. Next, it presents the architecture of the developed system and discusses some implementation issues. Then, it provides the results of performance evaluation of the PMRF.

It is our sincere hope that this volume provides stimulation and inspiration and that it will be used as a foundation for works to come.

June 2016

Yeong-Tae Song
Towson University, USA

Bixin Li
Southeast University, China

Contents

Contributors

E. Allison Newcomb Towson University, Towson, MD, USA

Saleh M. Alnaeli Department of Computer Science, University of Wisconsin-Fox Valley, Menasha, WI, USA

Satoshi Ashida Yamaguchi University, Yamaguchi, Japan

Afef Bahri MIRACL Laboratory, Higher School of Computing and Multimedia, Sfax, Tunisia

Kamel Barkaoui CEDRIC-CNAM, Paris Cedex 03, France

Meriem Bensouyad MISC Laboratory, A. Mehri, Constantine 2 University, Constantine, Algeria

Salem Chakhar Portsmouth Business School and Centre for Operational Research and Logistics, University of Portsmouth, Portsmouth, UK

Jinhyuck Choi Department of Computer and Information Sciences, Towson University, Towson, MD, USA

Djaouida Dahmani MOVEP, Computer Science Department, USTHB, Algiers, Algeria

Harry S. Delugach Department of Computer Science, University of Alabama in Huntsville, Huntsville, AL, USA

Weichao Gao Department of Computer and Information Systems, Towson University, Towson, MD, USA

Fatma Ezzahra Gmati RIADI Research Laboratory, National School of Computer Sciences, University of Manouba, Manouba, Tunisia

Nousseiba Guidoum MISC Laboratory, A. Mehri, Constantine 2 University, Constantine, Algeria

Robert J. Hammell II Department of Computer and Information Sciences, Towson University, Towson, MD, USA

Youngsub Han Department of Computer and Information Sciences, Towson University, Towson, MD, USA

Amirshahram Hematian Department of Computer and Information Sciences, Towson University, Towson, MD, USA

Xiaohui Hu Institute of Software Chinese Academy of Sciences, Beijing, China

Malika Ioualalen MOVEP, Computer Science Department, USTHB, Algiers, Algeria

Alessio Ishizaka Portsmouth Business School and Centre for Operational Research and Logistics, University of Portsmouth, Portsmouth, UK

Ikhyeon Jang Department of Information and Communication Engineering, Dongguk University Gyeongju, Gyeongbuk, South Korea

Ahmed Kheldoun MOVEP, Computer Science Department, USTHB, Algiers, Algeria; Sciences and Technology Faculty, Yahia Fares University, Medea, Algeria

Yanggon Kim Department of Computer and Information Sciences, Towson University, Towson, MD, USA

Daniel Ku US Army CECOM Communications-Electronics Research, Development and Engineering Center (CERDEC), Fort Sill, USA

Chao Lu Department of Computer and Information Sciences, Towson University, Towson, MD, USA

Mitsuho Matsumoto Tottori University, Tottori-shi, Japan

James Nguyen US Army CECOM Communications-Electronics Research, Development and Engineering Center (CERDEC), Fort Sill, USA

Djamel-Eddine Saïdouni MISC Laboratory, A. Mehri, Constantine 2 University, Constantine, Algeria

Philip Samuel Division of IT, School of Engineering, Cochin University of Science & Technology, Kochi, India

Xudong Sui Institute of Software Chinese Academy of Sciences, Beijing, China

E.V. Sunitha Department of Computer Science, Cochin University of Science & Technology, Kochi, India

Amanda Ali. Taha Department of Computer Science, University of Wisconsin-Fox Valley, Menasha, WI, USA

Yoshinobu Tamura Yamaguchi University, Yamaguchi, Japan

Tyler Timm Department of Computer Science, University of Wisconsin-Fox Valley, Menasha, WI, USA

Bingyang Wei Department of Computer Science, Midwestern State University, Wichita Falls, TX, USA

Nadia Yacoubi-Ayadi RIADI Research Laboratory, National School of Computer Sciences, University of Manouba, Manouba, Tunisia

Shigeru Yamada Tottori University, Tottori-shi, Japan

Yinan Yang Department of Computer and Information Sciences, Towson University, Towson, MD, USA

Sepideh Yazdani Japan International Institute of Technology (MJIIT), Universiti Teknologi Malaysia, Kuala Lumpur, Malaysia

Wei Yu Department of Computer and Information Systems, Towson University, Towson, MD, USA

Hanlin Zhang Department of Computer and Information Systems, Towson University, Towson, MD, USA

Jinfang Zhang Institute of Software Chinese Academy of Sciences, Beijing, China

Lei Zhang School of Computer Science, Beijing Information Science & Technology University, Beijing, China

Human Motion Analysis and Classification Using Radar Micro-Doppler Signatures

Amirshahram Hematian, Yinan Yang, Chao Lu and Sepideh Yazdani

Abstract The ability to detect and analyze micro motions in human body is a crucial task in surveillance systems. Although video based systems are currently available to address this problem, but they need high computational resources and under good environmental lighting condition to capture high quality images. In this paper, a novel non-parametric method is presented to detect and calculate human gait speed while analyzing human micro motions based on radar micro-Doppler signatures to classify human motions. The analysis was applied to real data captured by 10 GHz radar from real human targets in a parking lot. Each individual was asked to perform different motions like walking, running, holding a bag while running, etc. The analysis of the gathered data revealed the human motion directions, number of steps taken per second, and whether the person is swinging arms while moving or not. Based on human motion structure and limitations, motion profile of each individual was recognizable to find the combinations between walking or running, and holding an object or swinging arms. We conclude that by adopting this method we can detect human motion profiles in radar based on micro motions of arms and legs in human body for surveillance applications in adverse weather conditions.

Keywords Human motion analysis · Micro-Doppler signature · Radar surveillance · Classification

A. Hematian (✉) · Y. Yang · C. Lu
Department of Computer and Information Sciences, Towson University,
7800 York Rd, Towson, MD 21252, USA
e-mail: ahemat1@students.towson.edu

C. Lu
e-mail: clu@towson.edu

S. Yazdani
Japan International Institute of Technology (MJIIT),
Universiti Teknologi Malaysia, 54100 Kuala Lumpur, Malaysia

R. Lee (ed.), *Software Engineering Research, Management and Applications*, Studies in Computational Intelligence 654,
DOI 10.1007/978-3-319-33903-0_1

1 Introduction

In radar systems, in order to measure the radial velocity of a moving target, the well-known Doppler effect is used to do such measurement. Yet in real world not all the structural components of a target follow a perfect linear motion. It usually involves complex motions, called micro-motion, such as acceleration, rotation or vibration. As stated by Chen [1], the micro-motions, and their induced micro-Doppler effects were introduced to characterize the movement of a target. Consequently, various ways to extract and interpret radar micro-Doppler signatures becomes an active research area specially to observe micro-Doppler signatures from real life samples such as vehicle engine vibration, aircraft propellers or helicopter blades rotation, flapping wings of birds, swinging arms and legs of a walking person, and even heart-beat or respiration of a human. Returned radar signals from legs of a walking person with swinging arms contain micro-Doppler signatures relevant to human motion structure. The micro-Doppler effect empowers us to determine the dynamic properties of an object and it offers a new way to analyze object signatures. In order to determine the movement and identify specific types of targets based on their motion dynamics, the micro-Doppler effect may be used. We have already tested this concept in our previous work using simulated data to classify targets with simple movements [2–5]. Human motion is a combination of articulated and flexible motions that is accomplished by complex movements of human body parts. Human motion recognition has attracted ample attention in computer vision to become an approach for human gait motion analysis that is mostly based on extraction of the motions from a sequence of images [6]. Unlike DNA or fingerprint that are unique for each person, human gait motion features may not be unique as they can be easily replicated by doing the very same movements, especially when a radar is used to extract the movement of human gait [7]. In contrast to visual perception that is very sensitive to range, speed, light condition, and complex background, the radar does not suffer from visual limitations and uses micro-Doppler signatures to extract target movements. Consequently, the radar is very well suited for tracking human gait motions where visual target observation is not possible. This research involved micro-Doppler signature analysis of experimental data previously captured [3] from various human targets with different motions. The data was collected using an X-band FMCW (frequency modulated continuous wave) radar at 10 GHz. The experimental data presented here include different motion combinations between legs and arms like walking, fast walking, jogging, and running with one or two arms swinging, or even without any arm swinging. Moreover, some similar motion data was captured at different angles in respect to radar LOS (line of sight). An algorithm is implemented and tested by the given data to classify different human motions. Both our motion analysis algorithm and the results of radar data analysis are reported here. In this paper we shed more light on human motion structure and how each body part motion can be recognized and classified in radar data by using micro-Doppler signatures.

2 Micro-Doppler Signature Extraction from Radar Signal

Human gait has been studied for years in different fields of science like physiotherapy, rehabilitation and sports. As previously mentioned, human gait micro-motion is one of typical examples of radar micro-Doppler signatures. Many studies on radar micro-Doppler signatures for human gait analysis have been conducted since 1998 [1]. After studying earlier works on micro-Doppler [1–15], we are using previously captured radar data by Yang and Lu [3] to develop new motion analysis algorithm in order to extract and classify human gain micro-Motions. The radar data was captured in an empty parking lot as seen in Fig. 1. In order to minimize multi-path propagation, this site was selected to perform data capturing on weekends and it is on top of the roof. As stated in [3] a total of 60 people participated in the experiment. Each of them performed four different movements, including walking, walking at fast speed, jogging, and running at fast speed, for a duration of 5 s and more. Some of the people performed additional movements, including leaping, crawling, walking and running at different aspect angle, to fully illustrate the characteristic of the micro-Doppler signatures of human motions. Figure 2 shows micro-Doppler signatures of a human subject walking, fast walking, jogging, fast running and leaping. The image is centered at Doppler frequency, not zero frequency, to make it easier to compare micro-Doppler features between these motions. The similarities with all these motions are obvious, with biggest return from the torso form the center line, and the legs/arms swing causes the side band at both sides of the center line. The differences with these motions are also visible; the biggest difference is the shape of the leg swing, with the differences of the phase between leg swing and arm swing. These differences make classifying different motions possible.

Fig. 1 Radar data capturing environment [3]

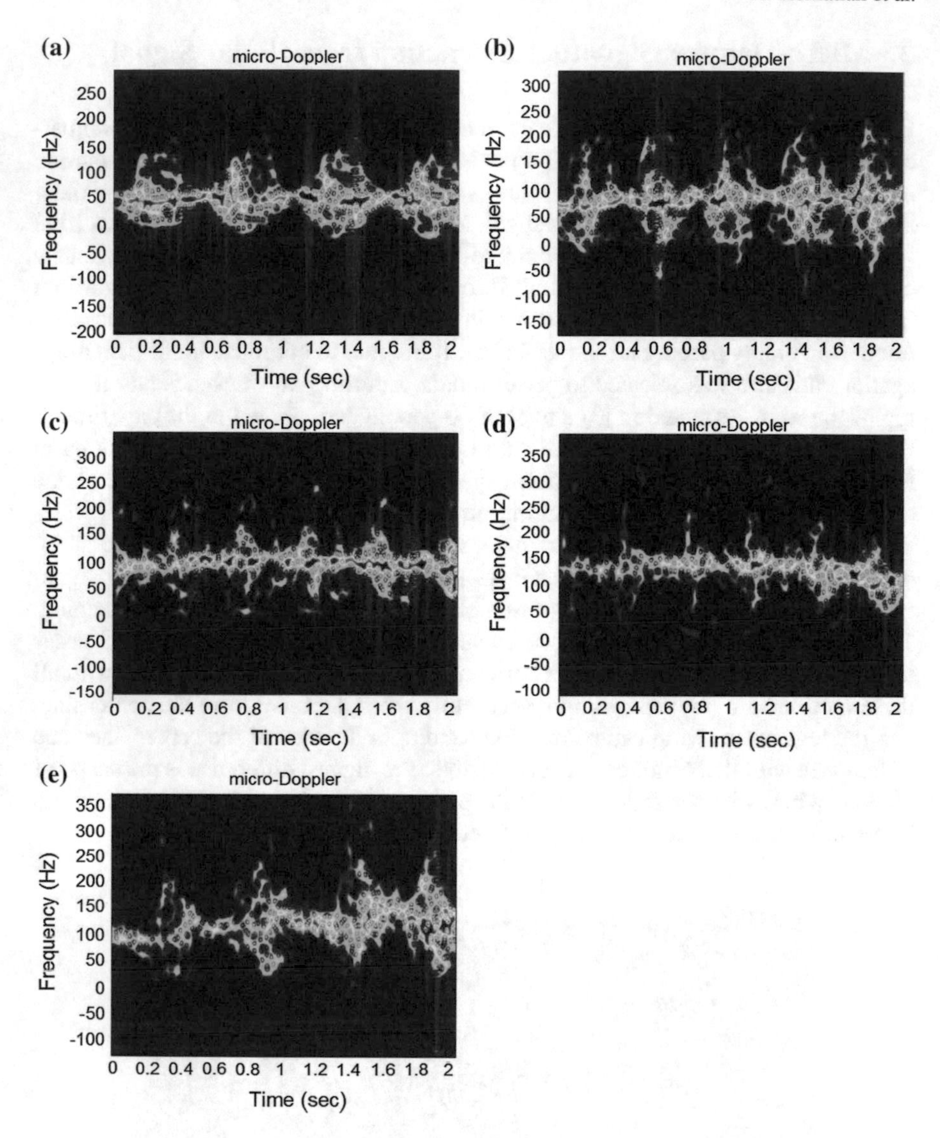

Fig. 2 Micro-Doppler signatures of different motions [3]. **a** Walking. **b** Fast Walking. **c** Jogging. **d** Fast running. **e** Sprinting

3 Human Gait Micro-motions Analysis

In this section, we explain how extracted micro-Doppler signatures are post-processed and analyzed to detect human body micro-Motions. These micro-Motions are compared against a simulated model of micro-Doppler signatures. Human micro-Motion classification results of captured radar data are also included in this paper.

Following is the overview of the proposed human motion analysis and classification using micro-Doppler signatures from radar data.

Radar Data Characteristics In order to start processing the captured data by a radar, we need to know the characteristics of the gathered data. The data we used [3] was captured at rate of 1256 samples per radar sweep. Radar sweep time was 1006 μs that brings our capture rate up to 994.0358 Hz at maximum distance resolution 128 cells. We used Matlab software to read and analyze the captured data. The given data files were in text format where all the content was numerical and readable. Each file was properly labeled with the information about the subject like: a unique file identification number, name of the person, motion types of legs and arms, extra objects used like a chair or bag and how they were used.

Extracting Doppler Signatures Extracting Doppler signatures from one dimensional data captured by radar is done by calculating Fast Furrier Transform (FFT) from each sweep of radar data. By putting the FFT data together we can easily distinguish between static and moving objects. Due to the existence of other objects in the capturing environment, we can clearly see static objects as straight horizontal lines and the moving subject as a diagonal curved line in Fig. 3a, as they are called the Doppler signatures of all objects. As indicated in Fig. 3a, the number of pulses recorded was 8956 pulses in 9.0097 s where the subject was running and swinging

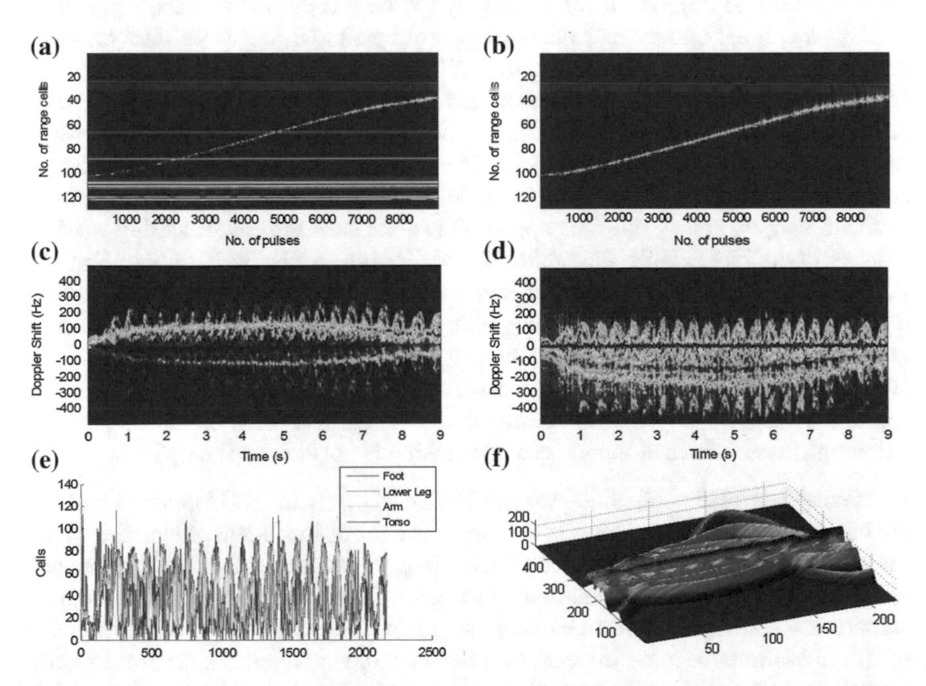

Fig. 3 Micro-motion analysis and classification of a running person. **a** Doppler signature (All objects). **b** Doppler signature (Moving objects). **c** Micro-Doppler signature. **d** Micro-Doppler signature without Doppler shift. **e** Body parts signatures. **f** Simulated 3D micro-Doppler signature

arms fast. Toward extracting the only moving objects from radar data, since the running person is moving from one cell to another, we can filter down the static objects by applying a high-pass filter on each cell. We decided to choose 60 Hz high-pass filter in order to suppress both static objects and electrical noises in the radar data. After extracting Doppler signatures from the radar data, we can calculate the average movement and determine the direction of the subject. In Fig. 3b, Doppler Signature of the moving person shows that our subject has moved from cell number 100–38 that the difference is −62 which indicates the moving subject is getting close to the radar. To extract the micro-Doppler signatures from Doppler signatures we need to determine a range of movement of the subject in the number of cells. Like here, the movement range is between cell numbers 38 and 100. By removing the rest of the data and only considering the data within the specified range, we can extract the micro-Doppler signatures of our targeted subject as shown in Fig. 3c.

Extracting Micro-Doppler Signatures The previous step of our algorithm only revealed the general motion of the moving subject. In order to extract the micro-Motions of the subject we need to convert the diagonal curved line into a straight line while preserving the details. To achieve this, we calculate the summation of each column of the data from previous step and converted our two dimensional data (cell-samples) into one dimension (samples). Then we apply one dimensional Gaussian filter to remove the noise and make the Doppler signal smoother as shown in Fig. 3d. Now it is time to use FFT again to display the frequency shifts of the summation of Doppler signal and extract micro-Doppler signatures. Since our data has only one dimension and we want to see the shifts of different frequencies, we need to have a sliding window to read the data and apply FFT iteratively. At each iteration starting from the first sample to 512th sample, we calculated the FFT that shows us the frequency changes up to 256 Hz (half of sample rate) and in next iteration we shifted the sliding window for one sample to calculate the FFT from 2nd sample to 513th sample, and so on. This operation is done until the end of sliding window reaches to the last sample available, for data shown in Fig. 3e is 8956th Sample. As shown in Fig. 3e, Micro-Doppler Signatures, it is clear that the person, who was running, started to accelerate at the beginning of the capturing process and decelerate at the end. This means, at this stage, the Doppler shift and micro-Doppler signatures are combined and the Doppler shift must be removed. Although in this 9 s of motion capture we can visually see the motion cycle of legs and arms of the subject, but extracting these information by signal processing is not easy as it may look.

Removing Doppler Shift Since we need to extract only micro-Doppler signatures for human micro-motion analysis, the Doppler shift of the captured signal must be removed. By removing the Doppler shift, the largest and main part of the human body which is torso will always remain at zero hertz. Consequently, the micro-Doppler signatures are always stabilized based on the torso of the target and repeated patterns of micro-Motions become clearer. In order to remove Doppler shift from micro-Doppler signatures that is a two dimensional data (Frequency-Time), we need to find the peak (maximum amplitude) of the frequency shifts from each column and then shift the data up or down within each column where the maximum peak meets the zero hertz frequency shift. Since we expect to find the maximum peaks from torso, a one dimensional median filter with window size 51 samples is applied to the index of the maximum peaks of all columns. This filter removes the unwanted noise from the list of indexes that may be caused by having the very same amplitude in few items close to shift frequency of torso at each column. Due to the fact that the human torso is the largest part of the human body, the maximum peak in the frequency shifts represents the torso where it reflects most of the radar signal. However, if the target moves toward the radar or gets far away from it, the amplitude of the frequency shifts increases or decreases, respectively. The major parameter that affects the reflected signal amplitude is the distance of the subject from the radar. The farther a subject moves the weaker signal we receive and we need to normalize the micro-Doppler signals in order to have consistent amplitudes for all micro-Motions and frequency shifts during a capture period.

Normalazing Micro-Doppler Signatures After removing the Doppler shift, we still see that depending of the general movement direction of the target, the micro-Doppler signatures get higher or lower amplitudes during time. In order to normalize the micro-Doppler signatures during time we decided to select the torso of human body as the reference point like the previous step. At this step we have a two dimensional data (Frequency-Time) to normalize. For each column, we find the maximum peak and calculate a multiplication ratio (MR). To calculate MR for each column, we need to find the largest peak among all maximum peaks of all columns to put as a reference amplitude and calculate the MRs. By finding the ratio between the maximum peak of each column and the largest peak of all, we calculate the MR and then multiply each column with its own MR. The normalized result is shown in Fig. 3d, Micro-Doppler Signatures Without Doppler Shift.

Extracting and Analysis of Micro-motion Signatures As shown in Fig. 3f (bottom right), the 3D simulated micro Doppler signatures of human body show that after torso, arms, lower legs and feet have consecutively lower amplitudes in normalized micro-Doppler signatures. Since, micro-Doppler signatures are always normalized by our method and human body mass variations and structure limitations cannot change radically during movement of the targets, we can define multiple thresholds in order to extract signatures for each part of the human body. Due to the symmetric motion of arms and legs and knowing that none of our targets have a lost arm or leg, we extracted the signatures for torso, both arms, both lower legs and both feet by using thresholding. Seeing that humans have many limitations like they cannot run faster than 45 km/h, we can draw a line around these limitations to calculate the average speed of a human target while running and define a threshold to distinguish walking from running based on the frequency of the steps taken per second. Same type of calculation can also be applied to human arms when swinging. We decided to extract the frequency of both feet and put two thresholds to determine the status of the subject. For the frequencies of higher than 2.5 Hz we consider that the target is running. For the frequencies between 2.5 and 1 Hz the target is walking and for lower frequencies the target is moving very slowly. Although human gait is based on motion of the legs, the rhythmic motions of the arms help the body to keep its balance. Thus, we expect to get the same frequencies from the signatures for arms and for the frequencies higher than 2.5 Hz we consider that the person is swinging both arms very fast. Between 2.5 and 1 Hz we consider that the person is swinging arms normally and for lower frequencies, if the person has a high gait speed then is

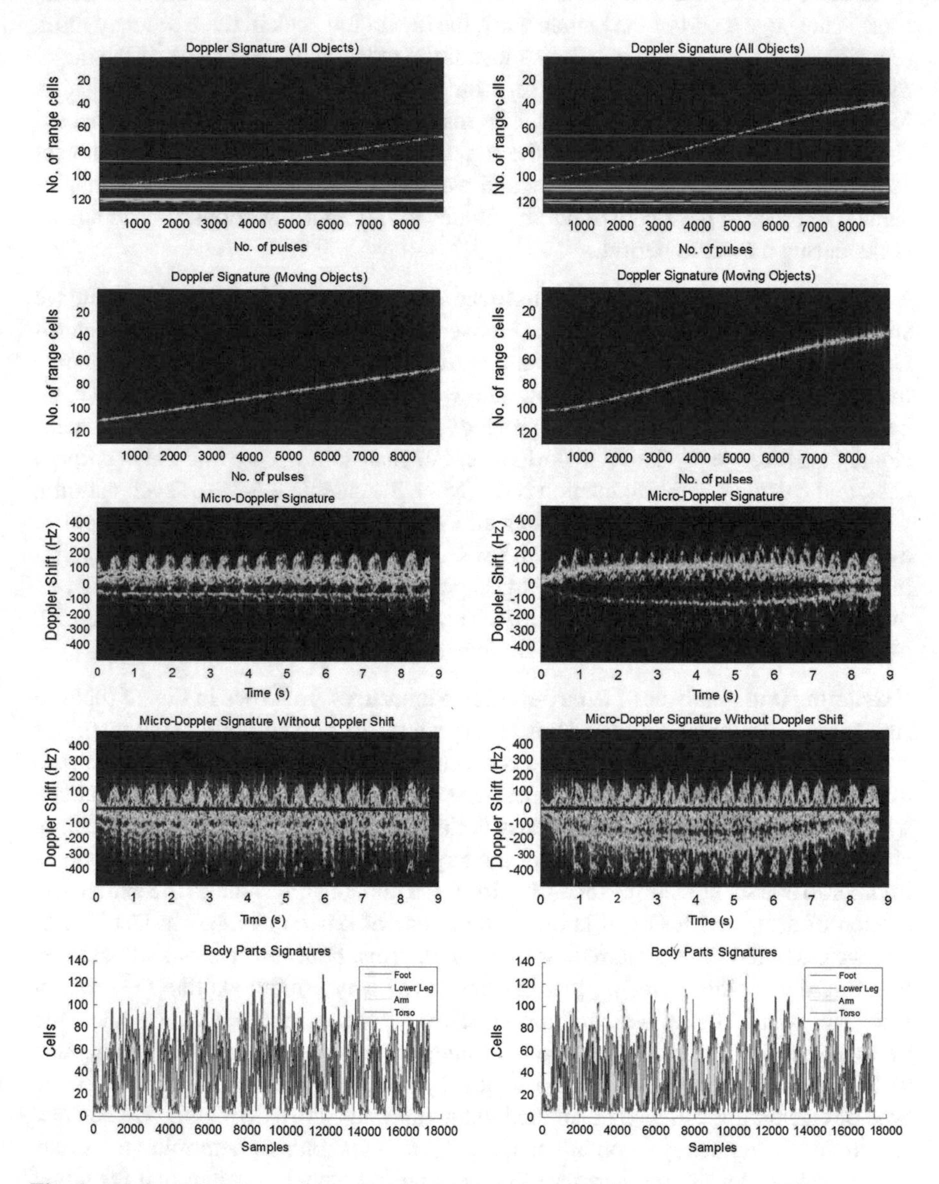

Fig. 4 Head to head comparison of micro-Doppler signatures between a walking (*Left column*) and running (*Right column*) person

holding something in his arm, otherwise he is not moving it. In Fig. 4, Body Parts Signatures, it is clear that a running person has a higher frequency than a walking person.

4 Human Gait Micro-motions Classification

Figure 4 shows the head to head comparison between each step of our method for typical running and walking states of a same person. As portrayed on the left column (walking), the target is walking almost at constant speed of 2.2198 steps per second toward radar while swinging arms. Unlike walking, the target starts to increase its speed at the beginning of the capture for running and decrease it at the end. On average, the target is running toward radar at speed of 2.6638 steps per second while swinging arms very fast. The mentioned human motion profiles are the most typical patterns of human micro-motions in normal situations that can be classified as normal behaviour of the target. In contrast to previous profiles, based on our experiments, a target that is running toward radar while holding a bag with both hands has a very different micro-Motion pattern. We calculated the average speed of 2.9971 steps per second while the average frequency of arms was only 0.39961 Hz that shows the target was on the move toward the radar while holding something in both arms. This type of human motion profile can be classified for triggering an alert system especially where radar is used for monitoring boarders, coasts or where visual observation is not possible.

5 Conclusion

In this paper, we proposed a new method for human motion analysis and classification based on micro-Doppler signatures in radar signal. The captured data characteristics and operating principle of our method including experimental results for each step have been demonstrated. It has been proved that using stabilized and normalized micro-Doppler signatures can reveal human micro-Motions and their patterns. By analyzing the micro-Motions of human body, we can distinguish and classify different motion profiles of a moving human target. Some motion profiles can be used as triggers for an alarm system in order to automatically reveal suspicious activities under radar surveillance. Such experiments with 2.4 GHz radar prove that micro-Doppler signatures carry ample information to detect and recognize human motions in radar signal. Consequently, by using radar data with higher sample rates and data resolution we will be able to extract more details of human micro-Motions like heart beat or respiration. Proposed method works based on offline radar data analysis and does not use real-time target tracking. In order to improve the accuracy and performance of our method for real-time radar applications we want to use FPGA to implement our method in VHDL. This can be considered as future work that radar target tracking and motion analysis information can be achieved in real-time for radar surveillance systems.

References

1. Chen, V. C. (2014). Advances in applications of radar micro-Doppler signatures. In *IEEE Conference on Antenna Measurements and Applications (CAMA)*, Conference Proceedings (pp. 1–4).
2. Lei, J., & Lu, C. (2005). Target classification based on micro-Doppler signatures. In *Radar Conference, IEEE International*, Conference Proceedings (pp. 179–183).
3. Yang, Y., & Lu, C. (2008). Human identifications using micro-Doppler signatures (pp. 69–73).
4. Yang, Y., Zhang, W., & Lu, C. (2008). Classify human motions using micro-Doppler radar. In *Conference Proceedings* (Vol. 6944, pp. 69 440V–69 440V–8). http://dx.doi.org/10.1117/12.779072.
5. Yang, Y., Lei, J., Zhang, W., & Lu, C. (2006). Target classification and pattern recognition using micro-Doppler radar signatures. In *Software Engineering, Artificial Intelligence, Networking, and Parallel/Distributed Computing. SNPD*, Conference Proceedings (pp. 213–217).
6. Zhang, Z., & Andreou, A. G. (2008). Human identification experiments using acoustic micro-Doppler signatures. In *Micro-Nanoelectronics, Technology and Applications. EAMTA*, Conference Proceedings (pp. 81–86).
7. Fairchild, D. P., & Narayanan, R. M. (2014). Classification of human motions using empirical mode decomposition of human micro-Doppler signatures. *Radar, Sonar and Navigation, IET*, *8*(5), 425–434.
8. Brewster, A., & Balleri, A. (2015). Extraction and analysis of micro-Doppler signatures by the empirical mode decomposition. In *Radar Conference (RadarCon), IEEE*, Conference Proceedings (pp. 0947–0951).
9. Chen, V. C. (2003). Micro-Doppler effect of micromotion dynamics: A review. In *AeroSense, International Society for Optics and Photonics*, Conference Proceedings (Vol. 5102, pp. 240–249). http://dx.doi.org/10.1117/12.488855.
10. Garreau, G., Andreou, C. M., Andreou, A. G., Georgiou, J., Dura-Bernal, S., Wennekers, T., & Denham, S. (2011). Gait-based person and gender recognition using micro-Doppler signatures. In *Biomedical Circuits and Systems Conference (BioCAS), IEEE*, Conference Proceedings (pp. 444–447).
11. Javier, R. J., & Youngwook, K. (2014). Application of linear predictive coding for human activity classification based on micro-Doppler signatures. *Geoscience and Remote Sensing Letters, IEEE*, *11*(10), 1831–1834.
12. Youngwook, K., & Hao, L. (2009). Human activity classification based on micro-Doppler signatures using a support vector machine. *IEEE Transactions on Geoscience and Remote Sensing*, *47*(5), 1328–1337.
13. Youngwook, K., Sungjae, H., & Jihoon, K. (2015). Human detection using Doppler radar based on physical characteristics of targets. *Geoscience and Remote Sensing Letters, IEEE*, *12*(2), 289–293.
14. Yang, Y., Qiu, Y., & Lu, C. (2005). Automatic target classification—experiments on the MSTAR SAR images. In *Software Engineering, Artificial Intelligence, Networking and Parallel/Distributed Computing, and First ACIS International Workshop on Self-Assembling Wireless Networks. SNPD/SAWN*, Conference Proceedings (pp. 2–7).
15. Wang, X., Li, J., Yang, Y., Lu, C., Kwan, C., & Ayhan, B. (2011). Comparison of three radars for through-the-wall sensing. In *Defense, Security, and Sensing Conference SPIE Proceedings*. SPIE, Conference Proceedings.

Performance Evaluation of NETCONF Protocol in MANET Using Emulation

Weichao Gao, James Nguyen, Daniel Ku, Hanlin Zhang and Wei Yu

Abstract The Mobile Ad-hoc Network (MANET) is an emerging infrastructure-free network constructed by self-organized mobile devices. In order to manage MANET, with its dynamic topology, several network management protocols have been proposed, and Network Configuration Protocol (NETCONF) is representative one. Nonetheless, the performance of these network management protocols on MANET remains unresolved. In this paper, we leverage the Common Open Research Emulator (CORE), a network emulation tool, to conduct the quantitative performance evaluation of NETCONF in an emulated MANET environment. We design a framework that captures the key characteristics of MANET (i.e., distance, mobility, and disruption), and develop subsequent emulation scenarios to perform the evaluation. Our experimental data illustrates how NETCONF performance is affected by each individual characteristic, and the results can serve as a guideline for deploying NETCONF in MANET.

Keywords NETCONF · YANG · MANET · Configuration management · Network management · Emulation

---

We would like to thank our Dr. Robert G. Cole for initially starting this effort and giving us feedback.

W. Gao (✉) · H. Zhang · W. Yu
Department of Computer and Information Systems, Towson University,
Towson, MD 21252, USA
e-mail: wgao3@students.towson.edu

H. Zhang
e-mail: hzhang4@students.towson.edu

W. Yu
e-mail: wyu@towson.edu

J. Nguyen · D. Ku
US Army CECOM Communications-Electronics Research,
Development and Engineering Center (CERDEC), Fort Sill, USA

R. Lee (ed.), *Software Engineering Research, Management and Applications*, Studies in Computational Intelligence 654,
DOI 10.1007/978-3-319-33903-0_2

1 Introduction

The emerging MANET has an infrastructure-free and dynamic network topology constructed by self-organized mobile devices. Compared to the traditional wired and access-point-based wireless network topologies, MANET has more flexible structures due to the frequent joining and exiting of nodes in the network. This flexibility, on one hand, enables MANET to better fit rapidly changing application scenarios, but on the other hand, increases the difficulty in managing the network.

Various network management protocols have been proposed, including NETCONF [1], Simple Network Management Protocol (SNMP) [2]. Although a large number of research efforts have focused on the performance evaluation of these protocols in wired networks, the performance of these management protocols on MANET are largely untested. In the research that does address this issue, various network simulation tools [3], such as NS-3 [4] and OMNET++ [5], were developed to evaluate the performance of protocols in MANET. However, these simulations provide the overall performance in the simulated topologies only, and the performance in real-world MANET topologies has not been thoroughly evaluated. In addition, while the simulated topologies were considered as the whole entities in the evaluation where the scale (i.e. number of nodes), random mobility pattern, and other parameters such as bandwidth and transmission range were present, the impacts of the individual characteristics of MANET on its performance are not clear.

In our investigation, we quantitatively evaluate the performance of NETCONF in the MANET environment. Unlike prior research that is primarily based on simulation, we leverage the network emulation tool, Common Open Research Emulator (CORE), to carry out the performance evaluation of NETCONF in MANET. To design the emulation scenarios, we develop a framework that captures the key characteristics of MANET (i.e., distance, mobility, and disruption). Specifically, distance encapsulates the structures that separate any two communicating nodes in MANET, mobility represents the leaving and joining of nodes, and disruption characterizes the environmental hindrance to the delivering packets over the network. Based on the designed framework, we develop a set of scenarios for the performance evaluation. The impact of distance is tested via increasing the number of hops in the scenarios where the MANET topologies are fixed. The impact of mobility is examined through varying the leaving time in the scenarios where a target node temporarily leaves the transmission range and losses connection. Finally, we introduce disruptions into the aforementioned scenarios to evaluate their impact.

We conduct extensive experiments to validate the performance of NETCONF in various scenarios. Our experimental data illustrates the performance of NETCONF with respect to individual characteristics and our data can be used as a guideline to deploy NETCONF in MANET. For example, the incremental hop can increase the delay in linear and exponentially enlarges the impact of disruption, and the disruption not only increases the delay for packet transmissions, but also reduce the capability of the NETCONF requests to tolerate the out-of-range time of a leaving node in the MANET topology.

The remainder of this paper is organized as follows: We conduct a literature review of the MANET and NETCONF and introduce the emulation tool CORE in Sect. 2. In Sect. 3, we describe the decomposition of MANET topology and the framework of simulation scenarios. In Sect. 4, we describe the testbed configuration and present the evaluation methodology and results. Finally, we conclude the paper in Sect. 5.

2 Background and Related Work

In this section, we provide the background of our work and conduct the literature review of areas relevant to our study.

MANET: MANET is a network with highly dynamic topologies that are constructed by mobile nodes. In a MANET, infrastructures such as access points are not required because nodes are self-organized and are able to act as routers in the network. These mobile nodes are able to move frequently and independently, and their neighboring nodes are continuously changing. The versatility of MANET makes it ideal for a variety of applications, including tactical networks, disaster recovery, entertainment, and the emerging Internet of Things (IoT) [6, 7].

With the advance of wireless technologies, there has been a tremendous increase in number of mobile nodes, leading to larger and more complex, topologies of the MANET. Many research efforts, for that reason, have investigated MANET in the following areas [7]: (i) Routing [8], (ii) Energy conservation [9], (iii) Quality of Service [10], (iv) Security [11], and (v) Network Management.

Several investigations have been devoted to studying network management protocols [12]. These efforts include the comparison of existing management protocols, and implanting these protocols in MANET environments. The former one focused on the performance comparison of protocols in the wired or wireless networks [13, 14]. For example, Slabicki and Grochla in [14] conducted the performance evaluation of SNMP, NETCONF and CWMP on a tree topology-based wireless network. The later one focused on how the existing protocols performed in a MANET environment [15, 16]. For example, Herberg et al. [15] evaluated the performance of SNMP in an Optimized Link State Routing Protocol (OLSRv2)-routed MANET environment.

NETCONF and YANG: NETCONF [1], which is an emerging network configuration protocol, is one of the network management protocols and its performance has been studied in wired or wireless networks based on access points [13, 14]. It was developed and standardized by the Internet Engineering Task Force (IETF) (in RFC6241), which provides the mechanism to retrieve, edit, and remove the configuration of devices within the network. NETCONF is a session-based protocol that enables multiple operations in the configuration procedures, and its client-server infrastructure over the secure transport (usually SSH over TCP) ensures the reliable and secure transactions.

The NETCONF protocol includes four layers of content, operations, messages, and secure transport. The content layer consists of the configuration data and notification data that is formed in XML. The operations layer defines the base protocol operations, such as < *get* >, < *edit* – *config* >, and < *delete* – *config* >, to retrieve and edit the configuration data. These operations and data are encoded in the massage layer as remote procedure calls or notifications. The messages are then transported over the secure transport layer between the client and server. YANG [17], the acronym for "Yet Another Next Generation", is the data modeling language for NETCONF protocol developed by IETF (in RFC 6020). It represents data structures in an XML tree format and is used to model the configuration and state data manipulated by the NETCONF protocol, NETCONF operations, and NETCONF notifications.

In the following, we describe the characteristics of NETCONF and YANG [18]: (i) *Session-oriented protocol over Secure Shell* (*SSH*): The NETCONF protocol consider SSH for its mandatory transport protocol. Other transports, including Transport Layer Security (TLS), Simple Object Access Protocol (SOAP), are considered to be optional. (ii) *Data-driven content with YANG*: The YANG is used to define the server API contract between the client and the server. (iii) *A stateful transaction model*: The NETCONF protocol is designed for one server and one client pair. The client and server exchange < *hello* > messages at the start of the session to determine the characteristics of the session that is always initiated by the client. Via it, the client then learns the exact data model content of YANG supported by the server. (iv) *RPC exchange messages encoded in XML*: A client encoded an RPC message in XML and sends it to a server. The server, in turn, responds with a reply, which is also encoded in XML. (v) *Database operations are standardized and extensible*: There are four operations used to manipulate the conceptual data defined in YANG: create, retrieve, update, and delete (CRUD). A datastore, which is a conceptual container with a well-defined transaction model, is defined to store and access information. The server advertises the datastores it supports in the form of capability strings.

CORE: Unlike the simulation-based approaches of other research efforts, we implement an emulation-based approach for the performance evaluation in this study. We utilize the Common Open Research Emulator (CORE) [19] to emulate the MANET environments. CORE is a python framework, providing a graphic user interface for building emulated networks. Consisting of an graphic user interface for drawing topologies of lightweight virtual machines (nodes) and Python modules for scripting network emulation, CORE is able to emulate a network environment (wired and/or wireless) and run applications and protocols on the emulated nodes. It also enables the connection between emulated networks and live networks. Using CORE, we are able to establish the scenarios for the performance evaluation of NETCONF in MANET environments.

3 Our Approach

In our data transaction model, we emulate the data traffic between one pair of nodes (source and destination), managed by NETCONF. We define "connected nodes" to be any pair of nodes, which are able to communicate in the MANET. Via deconstructing the complex MANET topology into several key characteristics, we formulate the test configurations to examine the individual and combined effects of these characteristics, and quantitatively measure their impact on the performance. In the following, we first present the key characteristics of MANET dynamic topology and then design the scenarios for emulation.

3.1 Key Characteristics

Figure 1 illustrates the coupling/decoupling model. Choosing n_1 and n_2 as a pair to observe the communication. Once n_2 moves to the range of n_4 as shown in the figure, n_1 and n_2 are "connected". During the data transmission between the connected pair, the performance will be impacted by the following three characteristics: (i) Distance, (ii) Mobility, and (iii) Disruption, which will be detailed in the next few subsections.

Distance: The distance between a pair of connected nodes is characterized by the combination of the physical and "processing" distance between them. It can be formalized in general terms as the sum of the delay of all of the hops on the route and the delay of all of the physical distances between each adjacent or connected node on the route. Figure 2 shows an example of the distance of a pair of nodes n_1 and n_2. The number of hops between n_1 and n_2 is 2 (n_3 and n_4) and the sum of physical distance between n_1 and n_2 is 700 m (250 + 250 + 200). The performance, as a measure of

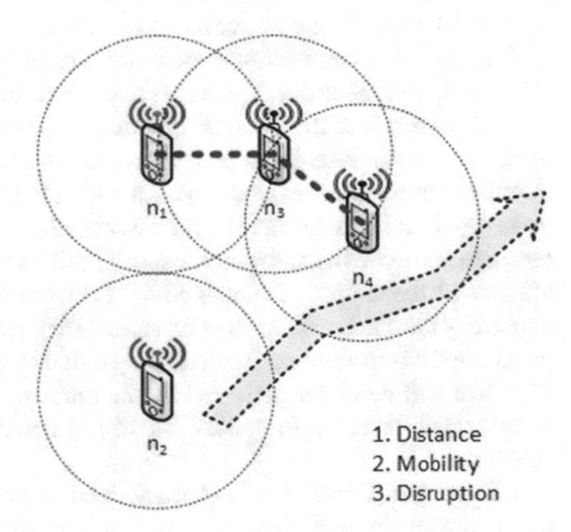

Fig. 1 MANET coupling/decoupling model

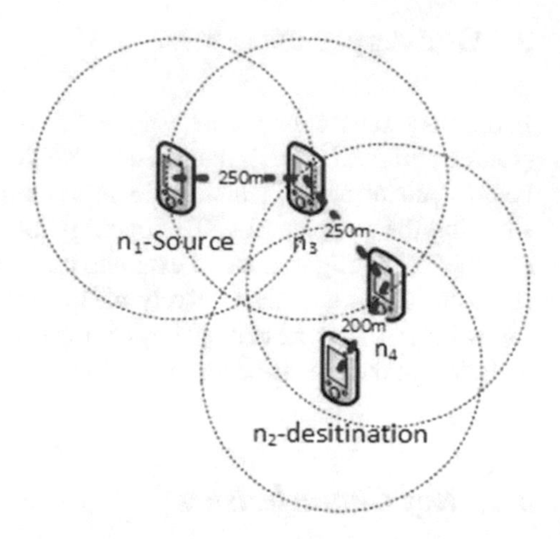

Fig. 2 Distance between two nodes (n_1 and n_2)

network speed or delay, is thus affected by the processing delay of every hop and the physical delay of data transmission over the distance. The latter can be computed by dividing the total physical distance by the speed of light. At any given point in time, both the distance and number of hops will be constants.

Mobility: This is the characteristic of the flexible morphology of the MANET. Mobility allows for the reconfiguration of the nodes, and therefore degrades performance due to the delay raised by disconnection and reconnection. The change in the connection between a pair of nodes can be modeled by three mobility patterns. We name the first pattern *Fixed Connection*, where the route does not change between the source and destination nodes and any of the nodes in the route during the data transmission. This pattern can be considered the same as a static topology. The second pattern is denoted *Same Route Returning*. During the data transmission, one or more nodes move out from the original route and cause the loss of connection between the source and destination nodes. The leaving nodes then returns and recovers the connection through the original route. The third pattern is named *Change Route Returning*. The nodes leave and cause the loss of connection as in the second pattern, but the difference is that the connection is reformed, either by the relocation of the remaining nodes, or after the leaving nodes return through a different route from the original one. This situation is considered because the routing table would be updated for the data transaction. Figure 3 illustrates the connection between two nodes n_1 and n_2 as the result of three mobility patterns. The performance affected by the mobility of nodes comes from the delay or failure of data transmission caused by the change of connection status. In the latter two patterns, the original data packet in the transmission is not delivered due to the loss of connection on the route. The data will be either delivered by the retransmission after the route is recovered or reformed, or failed to deliver due to a timeout, according to the retransmission schedule.

Disruption: Disruption is characterized as packet loss. There are many factors that can contribute to disruption, such as weak signal and shielding materials, and

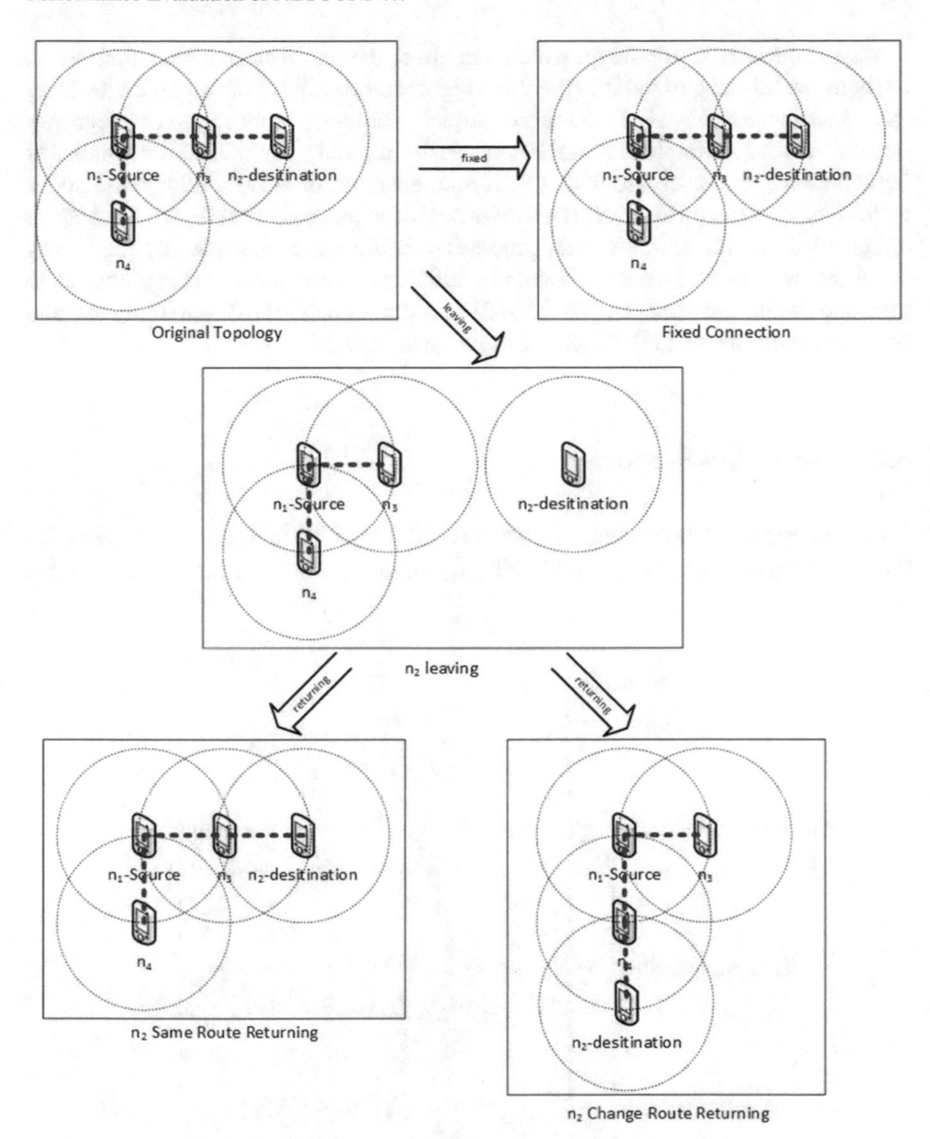

Fig. 3 Mobility patterns of nodes

in general these factors are the result of unintended or unavoidable environmental factors. The disruption results in the packet loss during the data transmission, which triggers the data retransmission and leads to additional delay, overhead, and sometimes the failure of the request or the response. The degree of disruption, with respect to the packet loss rate, can affect the performance of network management.

In a real-world MANET environment, the performance of NETCONF is affected by the complexity of the topologies. In our assumption, there are a variety of factors in a typical MANET topology, which can affect the performance. Nonetheless, all of them can be categorized into the three key characteristics. For example, the faster the nodes move in MANET, the more frequent the loss of connection and recovery occurs. Another example is that a larger density of nodes can reduce the number of hops between a pair of nodes that is communicating, because it allows the nodes to find shorter routes to reach other nodes. Under this assumption, we are able to create a 3-dimensional framework, generating the emulation scenarios by combining different levels of performance impact in individual dimensions. It also allows us to estimate the performance of NETCONF in a complex MANET environment, and find the conditions that NETCONF can perform as well.

3.2 Emulation Scenarios

To design the emulation scenarios, we create a 3-dimensional framework based on the key characteristics to capture MANET dynamic topology. Figure 4 illustrates this framework.

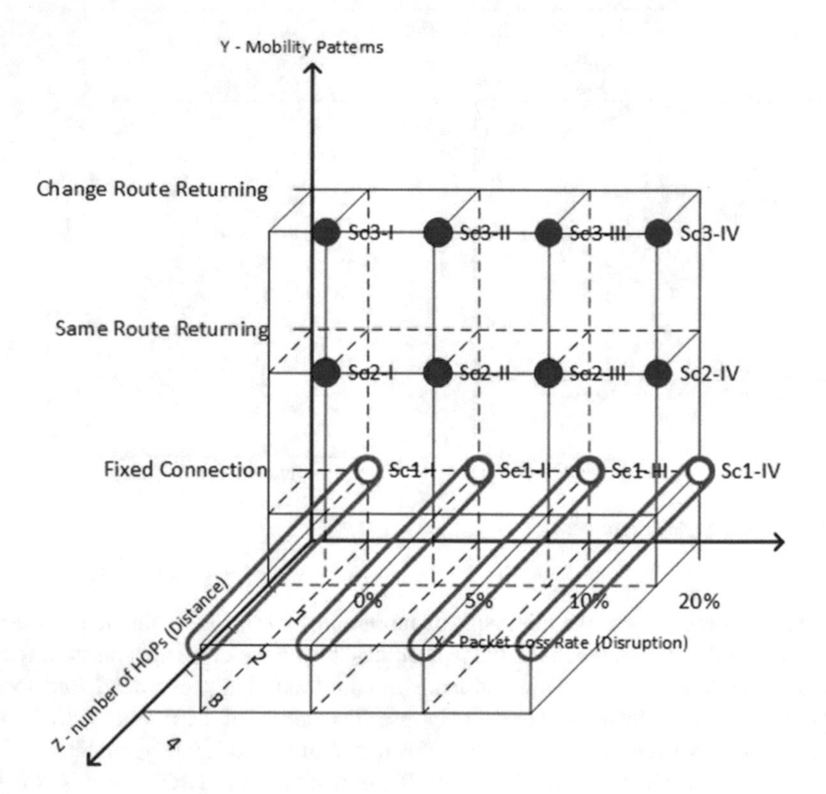

Fig. 4 Framework to design emulation scenarios

		Packet Loss Rate (Disruption)				Variables	Remark
		0%	5%	10%	20%		
Mobility Patterns	Fixed Connection	Sc1-I	Sc1-II	Sc1-III	Sc1-IV	HOPs	
	Same Route Returning	Sc2-I	Sc2-II	Sc2-III	Sc2-IV	Stay-out-time	HOP = 1
	Change Route Returning	Sc3-I	Sc3-II	Sc3-III	Sc3-IV	Stay-out-time	HOP = 1

Fig. 5 Scenario groups

The *X*-axis represents the dimension of *Disruption*, in the form of *Packet Loss Rate*. We evaluate four levels of packet loss rate, 0, 5, 10, and 20 %, to describe the degree of disruption to the MANET. The *Y*-axis represents the dimension of *Mobility*. In our experiment, we set only the destination node of the communicating pair as the candidate node to move, while all other nodes on the route are fixed. As illustrated in Fig. 3, the three mobility patterns are (i) *Fixed Connection*, where the destination node stays static during the data transmission, (ii) *Same Route Returning*, where the node leaves and returns to the previous position to recover the original route to the source node, and (iii) *Change Route Returning*, where the node leaves and returns to a new position and forms a new route. Notice that the duration that a node spends out of the network range can affect the performance, and therefore becomes a variable in our evaluation. The *Z*-axis represents the *Distance* between the pair of communicating nodes, in the form of number of hops. We set the distance between every two adjacent nodes on the route to an equal and fixed value such that the sum of physical distance is a constant multiplied by the number of hops plus one. The number of hops ranges from 0 to 4 in our emulation scenarios. By applying different level in each dimension, we generate 12 emulation scenarios grouped in 3 groups, as shown in Fig. 5.

- **Scenario Group 1** (*Sc1-I, Sc1-II, Sc1-III, and Sc1-IV*): Scenario Group 1 includes 4 subsets that represent the fixed MANET topologies, where the nodes are static during the data transmission. Each subset is evaluated with one of the four levels of packet loss rate as defined in the dimension of disruption, i.e. 0 % in $Sc1 - I$, 5 % in $Sc1 - II$, 10 % in $Sc1 - III$, and 20 % in $Sc1 - IV$. Additionally, each subset is evaluated at values of the variable hops 0 to 4. The goal is to observe how the distance affect the performance in a fixed topology under different levels of disruptions.
- **Scenario Group 2** (*Sc2-I, Sc2-II, Sc2-III, and Sc2-IV*): Scenario Group 2 includes 4 subsets that represent *Same Route Returning*, where the destination node moves out of the transmission range in the MANET, stays out of range for a variable period of time, and returns with the original route recovered. Each subset in this group is tested with the disruption levels described above. For the dimension of distance, we only consider the situation, where the number of hops is 1. This simplifies the scenario and provides the same structure in comparison with the

Change Route Returning scenario. The variable in each subset is the time that the leaving node stays out, named *stay-out-time*. We observe the delay and the success rate by increasing the *stay-out-time* in each emulation subset to evaluate the performance in *Same Route Returning* pattern across different levels of disruption. Another important assumption is that the leaving nodes "jump" out of the range instead of "move", which omits the time that the leaving nodes moves from the original position to the edge of the signal range. This allows us to evaluate the reconnection delay purely, without the contribution of the delay from varying distance.

- **Scenario Group 3** (*Sc3-I, Sc3-II, Sc3-III, and Sc3-IV*): Scenario Group 3 includes 4 subsets that represent the *Change Route Returning*, where the destination node moves out of the MANET range, stays out of range for a variable period of time, and returns with a new route formed. Just as in Scenario Group 2, we observe the delay and the success rate by increasing the *stay-out-time* in each subset, and evaluate the comparative performance in *Change Route Returning* under different levels of disruption.

By comparing the emulation results across the scenario groups, we can evaluate the impact of the mobility patterns on the performance of NETCONF. We can also study the impact of the disruption on the performance of network management protocol, by comparing the results across the four subsets in each scenario group.

4 Performance Evaluation

We setup an emulation environment to evaluate the performance of NETCONF based on scenarios designed in Sect. 3. The testbed utilizes CORE v4.8 to emulate the MANET environment. CORE requires a Linux environment, for which we used Ubuntu (v14.04 LTS). OpenYuma [20], which is an open source NETCONF implementation, contains a *netconfd* service for the server node and a *yangcli* client for the client node. Other necessary protocols and services in the emulation environment include OLSRv2 [21] for the MANET topology and SSH for NETCONF to run. We used olsrd2 [22] from the OLSR.org Network Framework (OONF) for the OLSRv2 protocol, and OpenSSH [23] for the SSH service.

The wireless connection of the emulated MANET environment is set to the default value of 54 Mbit/s, with the maximum range of 275 m. During the emulation, data traffic and the mobility script are set to start after the initial MANET topology is constructed. *Tcpdump* is also initialized to capture all TCP traffic, and the trace file is processed by using *awk* scripts for data analysis.

Several customized settings were required to override the default CORE configuration: (i) The OLSRv2 protocol is used in each emulated node with the OONF version. (ii) The SSH is directed the configuration file to a customized file that enables the ports and service for the netconfd server. A global key pair is setup for non-password login between emulated nodes. (iii) The service that launches the tcpdump

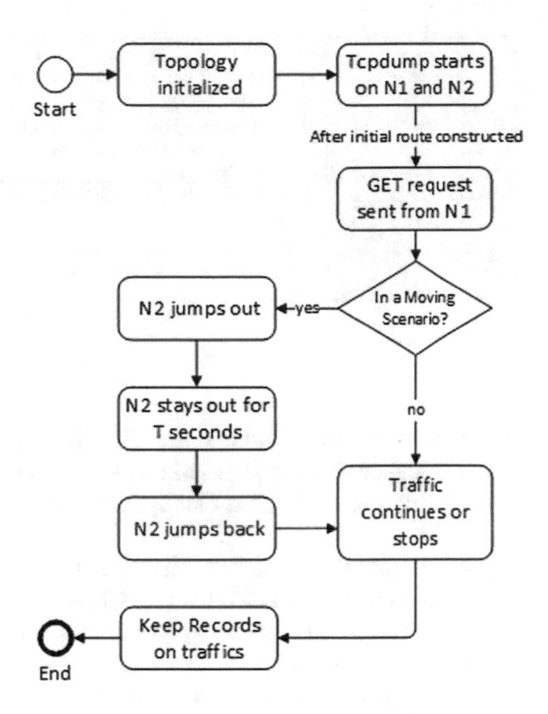

Fig. 6 Emulation process

for both server (n_1) and client (n_2) nodes is added and the request sending command on the client node is added. This configuration can be implemented by either customizing the node configuration in CORE's graphic user interface, or adding it to the Python script.

The quantitative evaluation of the performance of NETCONF is conducted through observing the traffic of simple NETCONF requests in the scenarios that we have created. The emulation in each scenario is designed as a simple NETCONF request and feedback process between a pair of target nodes in the MANET environment. We define the target pair of nodes named n_1 and n_2 for each scenario emulated in CORE. The node n_1 is set as the NETCONF client running *yangcli*, and the node n_2 is set as the NETCONF server running *netconfd*. The data traffic is a simple, one-time NETCONF server login and "GET" request sent from n_1 to n_2, and the expected feedback of the request from n_2. All TCP packets sent and received by n_1 and n_2 are recorded by *Tcpdump*.

Figure 6 illustrates the process of each emulation. When the emulation process starts, the MANET topology is initialized and constructed by the pair of target nodes along with the other nodes in between. *Tcpdump* starts to trace and record the TCP packets on both client (n_1) and server (n_2) nodes. Next, a "GET" command with login request, which is delay-triggered to wait for the construction of the MANET, is sent from the client (n_1). If it is in a moving scenario, the mobility script is triggered right after the request, and the server (n_2) starts jumping out from the current topology. The leaving node (n_2) stays out of range for T seconds, and moves back in range.

Metrics	Definition	Measurement
Packet Overhead	The difference between actual packet sent and packet need to send.	Extra packets sent due to the failure caused by any influencers
Largest Delay	The largest time interval of any packet from sending out to receipt of the acknowledgement	The total delay from the leaving of the nodes to the success of next packet transmission.
Recovery Delay	The difference between the largest delay and stay-out-time	The delay for topology reconstruction and packet retransmission.
Threshold of Transaction Failure	The stay-out-time that transactions start to fail	The tolerance of the leaving time at certain disruption level

Fig. 7 Metrics

The data traffic may continue after MANET is reconstructed or stops if the node stays out longer than the threshold. If it is a fixed topology, the server node stays in range during the emulation. Finally, the emulation stops and records everything to a trace file.

By repeating the emulation, changing the value of variables in each subset, and analyzing each trace record, we are able to obtain (i) the actual packets and bytes sent and received by each node, (ii) the number of packets and bytes needed to be sent for a connection, by the sequence number, (iii) the largest delay in each full request/feedback process, which is the largest time interval of any packet from being sent out to being acknowledged, and (iv) the stay-out-time that we define as the time that the leaving node stays out of range. To measure the performance of the NETCONF, we define the metrics in detail shown in Fig. 7. Each emulation scenario with be run 5 times.

Our evaluation is separated into the categories based on the two variables: distance and mobility. The third variable, disruption, is a variability condition for the evaluation of the distance and mobility. We will discuss the disruption when we show the results of scenarios associated with distance and mobility. Additionally, in our emulation we note that the NETCONF request can only be initialized successfully while the client and server nodes are connected. Furthermore, for any node that has left, the request can only be successfully delivered and executed once the route between the pair of nodes is recovered or reformed and the routing table is corrected, before timeout.

Distance: Figure 8 shows the variation of the largest delay from the number of hops in the Fixed Connection scenarios without disruption (Sc1-I defined in Sect. 5). It can be observed from the scatter chart that the largest delay (*Y*-Axis) increases linearly with the number of hops (*X*-Axis) between the client and server nodes, indicating a constant increase in delay per hop. Therefore, a steady route can be considered similar to the wired network or wireless network based on access points. Notice that the delay due to the physical distance between two nodes, at less than 1 μs for the transmission across 250 m, can be discarded because the total delay is in the range of 100 ms, leaving us to simply the delay for the hops.

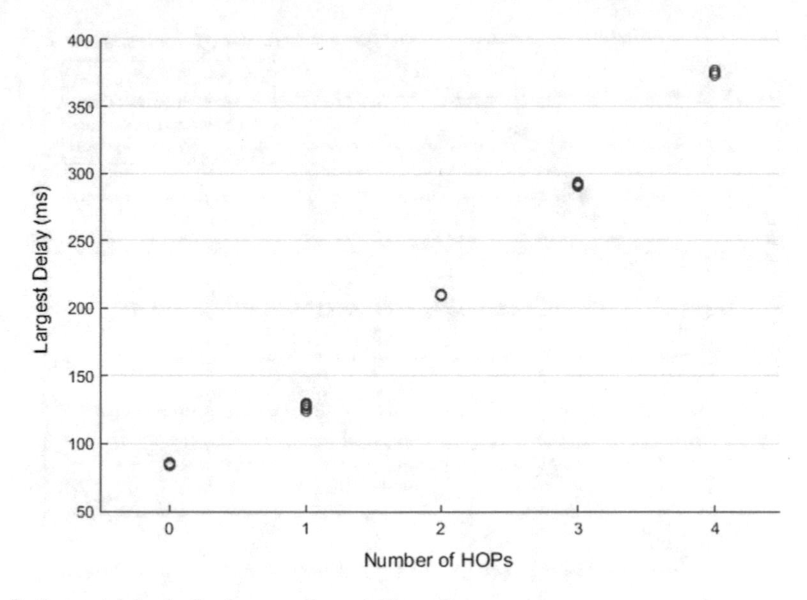

Fig. 8 Largest delay in fixed connection w/o disruption

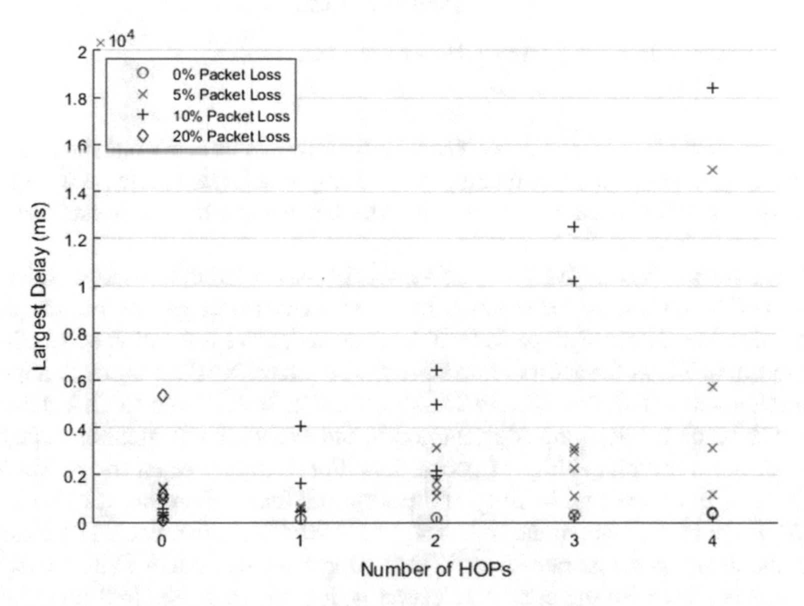

Fig. 9 Largest delay in fixed connection w/disruptions

When disruption occurs, the delay increases with increasing the number of hops. Figure 9 compares the distribution of largest delay by hops under 4 different levels of disruptions (results of the emulations in Group 1: Sc1-I, Sc1-II, Sc1-III, and Sc1-IV defined in Sect. 5). Under a higher disruption level, as observed in the scatter chart,

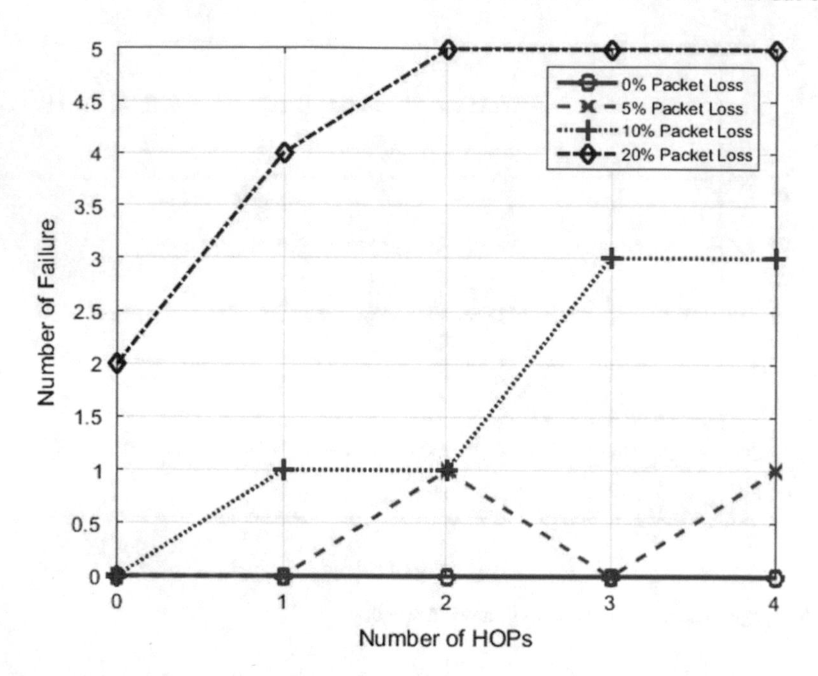

Fig. 10 Transaction failure in fixed connection w/disruption

the upper bound of the largest delay increases significantly by every incremental hop. For example, under a 10 % packet loss rate, the largest delay can increase from 1500 ms at 0 hop to 15000 ms at four hops, where the delay under the 0 % packet loss rate increases only from 86 to 375 ms.

In Fig. 10, we directly compare the number of packet failures by hops under the 4 levels of disruption. In a disruption free environment (0 % packet loss rate), the request does not fail since all packet will be delivered. Once disruption occurs, however, the probability of request failure increases by increasing the number of hops. In these scenarios (Sc1-II, Sc1-III, and Sc1-IV defined in Sect. 5), a packet is lost during the transmission between the adjacent nodes, and every incremental hop exponentially increases the probability of packet loss. For example, when the packet loss rate is 10 %, the overall probability of a packet loss for a 2-hop route is 27.1 % (or $1 - 90\,\%^3$) and for a 3-hop route is 34.4 % (or $1 - 90\,\%^4$). As the scatter chart shows, under the 10 % packet loss rate, the NETCONF request fails 1 out of 5 attempts when there is 1 hop between the client and server nodes, and increases to 3 times when number of hops raises to 3.

Once the packet loss triggers the retransmission, the delivery of the packet is delayed, and a greater packet overhead is incurred. Figure 11 compares the number of duplicate packets (overhead) by hops for each level of disruption. In each scenario where the packet loss occurs (Sc1-II, Sc1-III, and Sc1-IV defined in Sect. 5), the overhead increases with increasing hops. This distribution pattern of Overhead-Distance matches the Delay-Distance, where the delay is a measure of the time effect

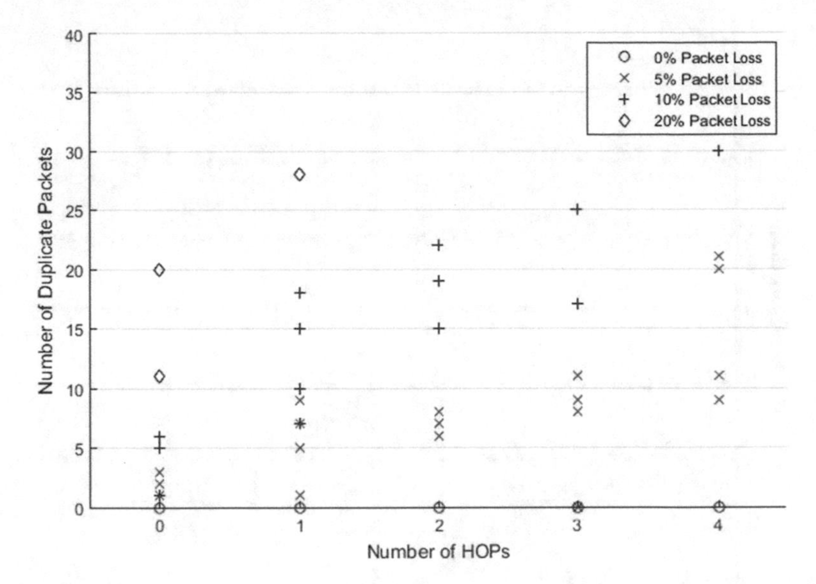

Fig. 11 Packet overhead in fixed connection

of retransmitting packets. Likewise, as the probability of disruption is exponential with distance, so is overhead, as each disruption results in a retransmission. It follows, then, that the Delay-Distance trend also increases exponentially.

The result of emulation in Scenario Group 1 indicates how the distance between two pairs of nodes will affect the performance of the NETCONF traffic in MANET. Greater distance linearly increases the delay in a disruption-free environment, and exponentially increases the delay and overhead proportionally with the probability of disruption.

Mobility: While Scenario Group 1 represents the performance of NETCONF traffic in the steady connection circumstances, Scenario Group 2 and 3 shows the influences resulting from the moving nodes. Excluding the disruption (Sc2-I defined in Sect. 5), Fig. 12 shows the variation of the *Recovery Delay* from the *stay-out-time* in the *Same-Route-Returning* scenarios without disruption (Sc-2-I defined in Sect. 5). Theoretically, the time for restructuring a same topology MANET should be steady. The scatter chart, however, indicates a non-linear pattern of *Recovery Delay* by stay-out-time. With the increase of stay-out-time, the delay jumps to a high level and slowly decreases repeatedly. This indicates that the retransmission may not be triggered right after the recovery of the route, and the time interval from the recovery of route to the actual retransmission varies by *stay-out-time*. When not deducting the *stay-out-time*, Fig. 13 shows the scatter chart of the *Largest Delay* (Y-Axis) by *stay-out-time* (X-Axis). It can be observed that the *Largest Delay* increases in almost double at certain time point and remains the same level until next jump. This is because the time a packet to be re-sent is scheduled after the last failure, once the leaving node moves back and recovers the route before next scheduled transmission

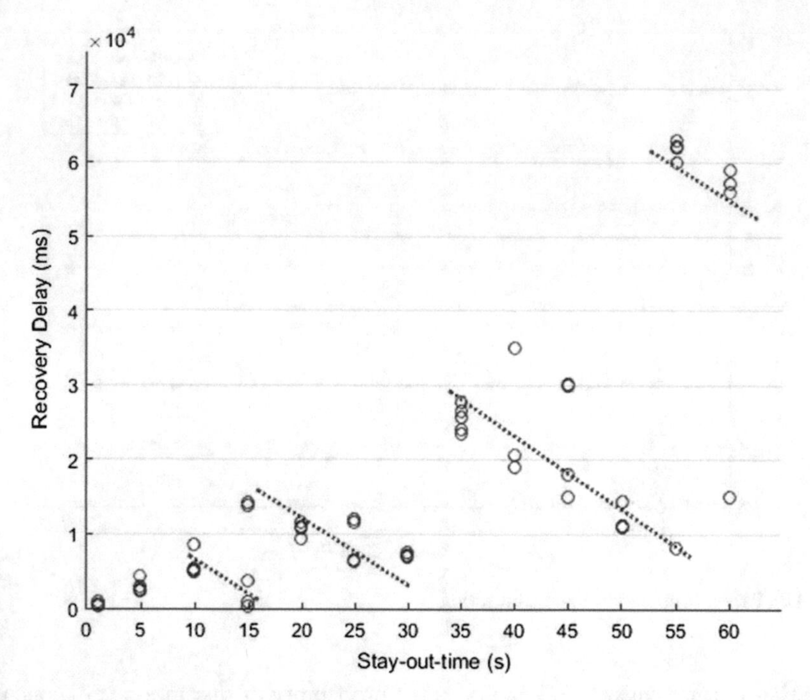

Fig. 12 Recovery delay in SameRouteReturning w/o disruption

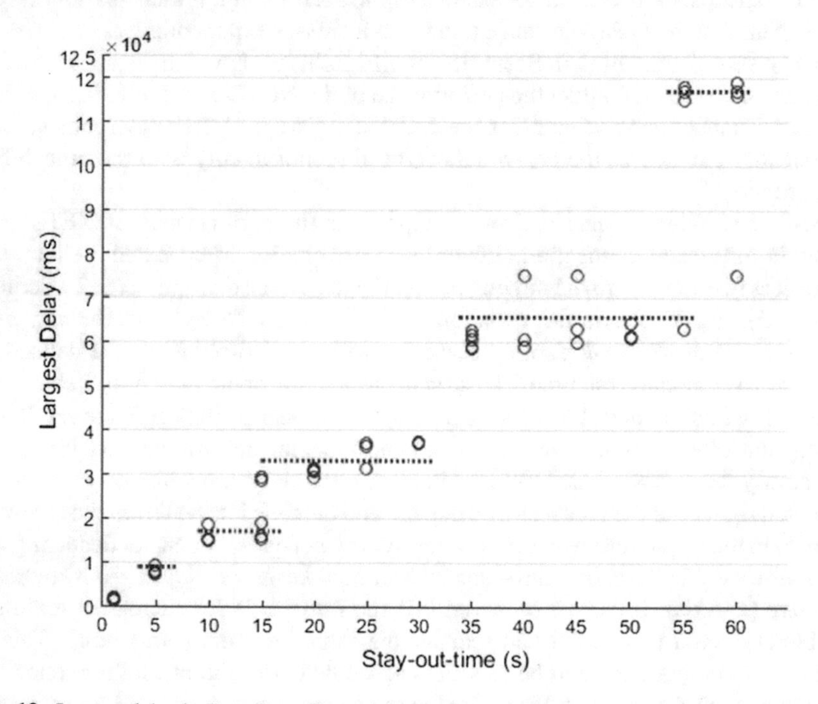

Fig. 13 Largest delay in SameRouteReturning w/o disruption

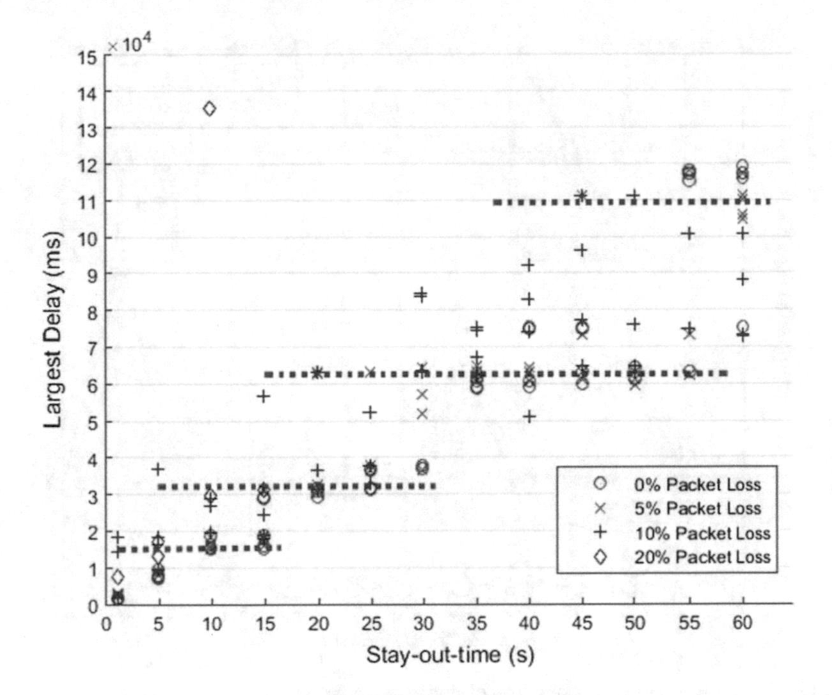

Fig. 14 Largest delay in SameRouteReturning w/disruption

time, it has to wait until the time is reached. Thus, earlier returning nodes would have the same retransmission time. For example, a node leaves for 40 and 50 s will both retransmit its packet at 68000 ms, leading to the same delay level in the figure.

Once disruption is taken into consideration (Sc-2-II, Sc-2-III, and Sc-2-IV defined in Sect. 5), the packet loss will increase the *Largest Delay* of a target *stay-out-time* to a higher level than the scenario without disruption, due to the failure of packet transmission. Figure 14 compares the *Largest Delay* by *Stay-out-time* in the *Same Route Returning* pattern across different levels of disruption. The figure shows that a higher disruption level like 10 % packet loss rate can increase the largest delay from 17000 to 32000 ms or even 65000 ms level for the same *stay-out-time* (15 s). Additionally, there are a significant number of data points that fall between two TCP retransmission schedule delay levels, due to the additional retransmission because of dropped packets. This increase of the largest delay by disruption also lowers the tolerance to the out-of-range time. Figure 15 compares the transaction failures in *Same Route Returning* among different levels of disruptions. For example, when the leaving node stays out for 10 s, there is 1 failure under 5 % packet loss rate scenario, 2 under 10 % scenario, and 3 under 20 % scenario, respectively. During the evaluation, we also find that, under the disruption-free scenario, the failure starts to occur at 66 s of stay-out-time.

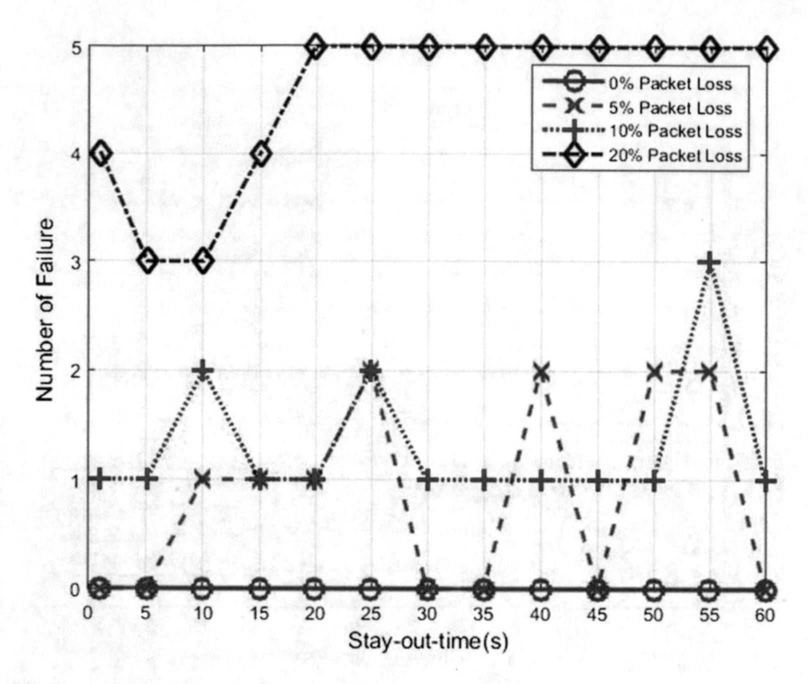

Fig. 15 Transaction failure in SameRouteReturning w/disruption

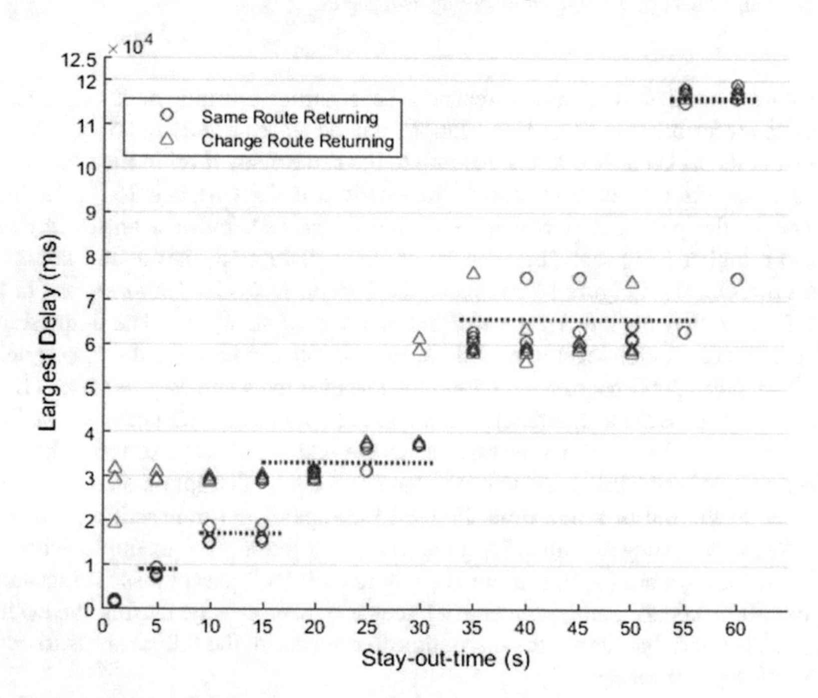

Fig. 16 Largest delay in Same versus ChangeRouteReturning w/o disruption

In the *Change Route Returning* Scenario without disruption (Sc-3-I defined in Sect. 5), the pattern is similar to the *Same Route Returning* with under the same conditions. In Fig. 16, we compare the *Largest Delay* by *stay-out-time* between *Same Route Returning* and *Change Route Returning*. It can be observed that when *stay-out-time* is above 15 s, the two scenarios follows the same pattern. Nonetheless, when the *stay-out-time* is under 15 s, the *Largest Delay* is a constant of 30 s for *Change*Route*Returning*. This contrasts the Same Route Returning, where there are *Largest Delay* minima at 10 and 20 s, for *stay-out-times* of 5 and 10–15 s respectively. The distinction can be explained by the time necessary to correct the routing table for the *Change Route Returning* that does not exist for the *Same Route Returning*.

When disruption is involved (Sc-3-II, Sc-3-III, and Sc-3-IV defined in Sect. 5), *Change Route Returning* follows the same pattern as in *Same Route Returning* scenarios. Figures 17 and 18 compare the largest delay and transaction failure in different levels under different levels of disruption. From these figures, the result of emulation in Scenario Groups 2 and 3 indicates how the mobility patten of the leaving node will affect the performance of the NETCONF traffic in a MANET topology. The NETCONF requests in a MANET environment rely on the TCP retransmission strategy. The delay caused by the change of route in a MANET is determined by the TCP retransmission scheduling. Requests can be successfully sent, received, and executed as long as the route can be reconstructed before the last retransmission try. In a disruption-free environment, the transaction can tolerate 60 s for a leaving node

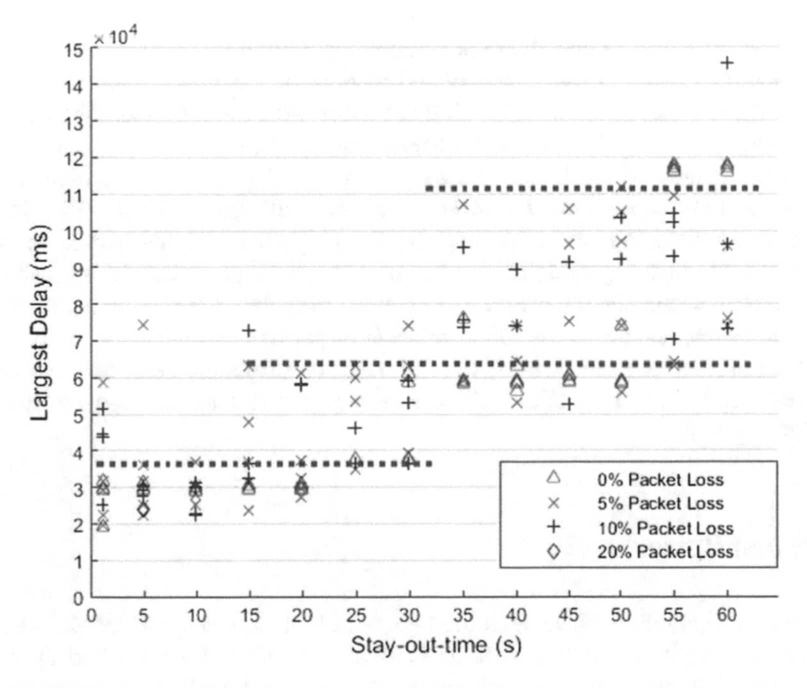

Fig. 17 Largest delay in ChangeRouteReturning w/disruption

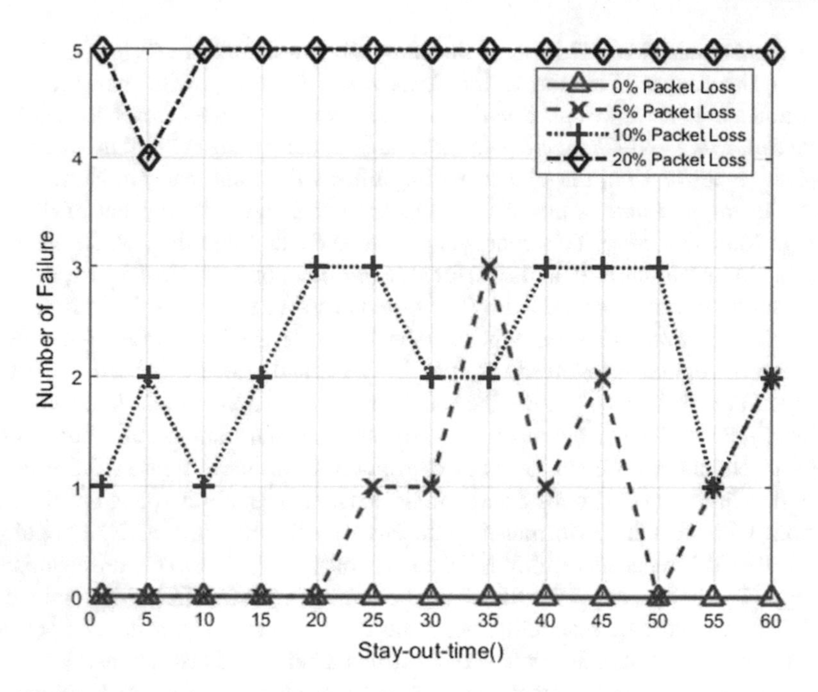

Fig. 18 Transaction failure in ChangeRouteReturning w/disruption

to stay out of range. Finally, there is no significant difference whether the original route is reconstructed or changed once the stay-out-time is greater than 15 s.

Disruption: Based on the emulation scenarios we have designed, the impact of disruption is compared among the 4 scenarios. Again, in Figs. 9, 14, and 17, the largest delay under 4° of disruption are compared in *Fixed Connection*, *Same Route Returning*, and *Change Route Returning* scenarios, and Figs. 10, 15, and 18 compare the request failure. As we can observe form each figure, a higher packet loss rate always results in a higher delay and the number of failures. This is because in a disruptive environment, the packet loss occurs during the transmission between two adjacent nodes, leading to the retransmission of packets. A higher packet loss rate can lead to more retransmissions, which further increases the delay and overhead, and reduces the capability of the NETCONF request to tolerate the *stay-out-time* of the leaving nodes.

5 Final Remarks

In our investigation, we leveraged the network emulation tool, CORE and carried out the quantitative evaluation of NETCONF in a MANET. We developed a generic framework that considers the key characteristics of MANET (distance, mobility, and disruption) and designed scenarios to perform the emulation study. Our experimental data show how NETCONF performance was affected by individual characteristics, and the results can serve as a guideline for deploying NETCONF in MANET. The 3-dimensional framework that we designed to create MANET emulation scenarios can be applied not only to evaluate NETCONF, but to other protocols that are applicable to the MANET topology.

References

1. Enns, R., Bjorklund, M., Schoenwaelder, J., & Bierman, A. (2011). Network configuration protocol (netconf). Internet Engineering Task Force, RFC 6241.
2. Harrington, D., Presuhn, R., & Wijnen, B. (2002). An architecture for describing simple network management protocol (snmp) management frameworks. Internet Engineering Task Force, RFC 3411.
3. Imran, M., Said, A., & Hasbullah, H. (2010). An overview of mobile ad hoc networks: Applications and challenges. In *Proceedings of 2010 International Symposium Information Technology (ITSim)*.
4. Network Simulator V3 (NS-3). http://www.nsnam.org.
5. OMNET++. http://www.omnetpp.org.
6. Bellavista, P., Cardone, G., Corradi, A., & Foschini, L. (2013). Convergence of manet and wsn in iot urban scenarios. *IEEE Sensor Journal, 13*(10), 3558–3567.
7. Hoebeke, J., Moerman, I., Dhoedt, B., & Demeester, P. (2004). An overview of mobile ad hoc networks: Applications and challenges. *Journal of the Communications Network (JCN), 3*(3), 60–66.
8. Patel, D. N., Patel, S. B., Kothadiya, H. R., Jethwa, P. D., & Jhaveri, R. H. (2014). A survey of reactive routing protocols in manet. In *Proceedings of 2014 IEEE International Conference on Information Communication and Embedded Systems (ICICES)*.
9. Chawda, K., Gorana, D. (2015). A survey of energy efficient routing protocol in manet. In *Proceedings of 2nd International Conference on Electronics and Communication Systems (ICECS)*.
10. Yu, Y., Ni, L., & Zheng, Y. (2011). Survey of qos multicast routing protocols in manets. In *Proceedings of 2011 International Conference on Computer Science and Service System (CSSS)*.
11. Alani, M. M. (2014). Manet security: A survey. In *Proceedings of 2014 IEEE International Conference on Control System, Computing and Engineering (ICCSCE)*.
12. Goncalves, P. (2009). An evaluation of network management protocols. In *Proceedings of IFIP/IEEE International Symposium on Integrated Network Management*.
13. Hedstrom, B., Watwe, A., & Sakthidharanr, S. (2011). http://morse.colorado.edu/~tlen5710/11s.
14. Slabicki, M., & Grochla, K. (2014). Performance evaluation of snmp, netconf and cwmp management protocols in wireless network. In *Proceedings of the 4th Internationa Conference on Electronics, Communications and Networks*.
15. Herberg, U., Cole, R. G., & Yi, J. (2011). Performance analysis of snmp in olsrv2-routed manets. In *Proceedings of the 7th International Conference on Network and Services Management, October 24–28*.
16. Kuthethoor, G. (2008). Performance improvements to netconf for airborne tactical networks. In *Proceedings of IEEE Intertional Conference on Military Communication (MILCOM)*.
17. Bjorklund, M. (2010). Yang—a data modeling language for the network configuration protocol (netconf). Internet Engineering Task Force, RFC 6020.
18. NETCONFCENTRAL. https://www.netconfcentral.org.
19. Common Open Research Emulator. http://www.nrl.navy.mil/itd/ncs/products/core.

20. OpenYuma. https://github.com/OpenClovis/OpenYuma.
21. Herberg, U., Clausen, T., Jacquet, P., & Dearlove, C. (2014). The optimized link state routing protocol version 2. Internet Engineering Task Force, RFC 7181.
22. OLSRD2. http://www.olsr.org.
23. OpenSSH. http://www.openssh.com.

A Fuzzy Logic Utility Framework (FLUF) to Support Information Assurance

E. Allison Newcomb and Robert J. Hammell II

Abstract The highly complex and dynamic nature of information and communications networks necessitates that cyber defenders make decisions under uncertainty within a time-constrained environment using incomplete information. There is an abundance of network security tools on the market; these products collect massive amounts of data, perform event correlations, and alert cyber defenders to potential problems. The real challenge is in making sense of the data, turning it into useful information, and acting upon it in time for it to be effective. This is known as *actionable knowledge*. This paper discusses the use of fuzzy logic for accelerating the transformation of network monitoring tool alerts to actionable knowledge, suggests process improvement that combines information assurance and cyber defender expertise for holistic computer network defense, and describes an experimental design for collecting empirical data to support the continued research in this area.

Keywords Fuzzy logic · Information assurance · Network defense · Decision support · Process improvement

1 Introduction

Our growing dependence on computers and information, paired with the ubiquity of computing devices, demands we treat cyber security as a suite of many different problems involving many different, often unpredictable, outcomes. Cyber defense is

E. Allison Newcomb (✉)
Towson University, Towson, MD 21252, USA
e-mail: enewco2@students.towson.edu

R.J. Hammell II
Department of Computer and Information Sciences,
Towson University, Towson, MD 21252, USA
e-mail: rhammell@towson.edu

R. Lee (ed.), *Software Engineering Research, Management and Applications*, Studies in Computational Intelligence 654,
DOI 10.1007/978-3-319-33903-0_3

performed in Computer Network Defense (CND) centers or Cyber Security Operations Centers (CSOC). Teams of cyber defenders working in these centers are composed of system administrators, network defense analysts and forensics experts. Cyber defenders must protect all access points, be able to quickly assess and minimize damage and/or downtime, and detect and eliminate adversarial presence in the network. These are monumental tasks when considering the dynamic nature of network assets, their potentially diverse configurations, and the fact than an adversary needs only to find one way into the infrastructure to cause problems.

Detecting, analyzing and responding to unusual network activities are critical functions performed by cyber defenders. There is an abundance of network monitoring tools, Intrusion Detection Systems (IDS), Intrusion Prevention Systems (IPS), and Security Information and Event Management (SIEM) products and services to integrate any or all of the outputs from those systems. The amount of data and information generated by these technologies is overwhelming. As an example, consider a recent article [1] that states more than 50 TB of data can be generated by a medium-sized enterprise network in a 24-h period. Assuming the software tools used by CND or CSOC were accurate and reliable, cyber defenders would be presented with approximately 5 GB of data to process each second to recognize cyber attacks in near real-time.

It is clear that cyber defenders have a deluge of information and would benefit from automation that assists in prioritizing that information. The real need is the ability to turn all this information into *actionable knowledge*—useful information that can be acted upon in time to provide effective cyber defense [2].

Vulnerability analyses on computer systems and information networks have been performed for well over a decade. So many vulnerabilities exist, with new ones emerging almost daily, that it is practically impossible to address them all. These analyses have historically taken a defensive posture. For a more complete understanding of weaknesses, perhaps we should think in terms of what the adversary might target rather than just performing vulnerability scans and running security readiness scripts; that is, take an offensive posture. The adage postulating that "the best defense is a good offense" suggests that we should know as much as possible about how an adversary might exploit, or target, a weakness.

The work presented here incorporates a targeting methodology developed for and used by US Special Forces units to determine the feasibility of a plan of attack. This targeting methodology has been adopted and modified for risk assessments to our nation's food supply, and vulnerability assessments for critical infrastructure. The methodology, CARVER (*Criticality, Accessibility, Recuperability, Vulnerability, Effect, Recognizability*), will be detailed in a later section of this paper.

The research we have undertaken seeks to help cyber defenders function more efficiently and lighten their burden of analysis and interpretation through a combination of a fuzzy-based decision support tool and business process improvement strategy. This paper introduces the Fuzzy Logic Utility Framework (FLUF); a decision support tool to assist cyber analysts in addressing intrusion detection alerts on devices deemed most critical relative to an organization's mission. The FLUF is used to determine the Alert Priority Rating (APR), which arranges IDS/IPS alerts in high to low order of severity. The FLUF considers the criticality of the affected asset, the level of access needed to infiltrate the asset, and the effect of the affected asset's compromise to the enterprise or mission.

The framework also supports process improvement by providing a means for the teams most familiar with the network architecture to participate in defining the rules that dictate alert priority and recommended actions. It is designed to be customized by individual organizations to address its particular needs and concerns. Further, because it is a framework, different components from the CARVER or other risk assessment methodology can be used in place of the components selected for this preliminary design. Elements of the Risk Management Framework (RMF) may also be used instead of the CARVER approach.

While presenting our approach for prioritizing network sensor alerts to provide improved support to cyber defenders, the work herein provides several main contributions. First, a fuzzy logic architecture previously constructed to assist military intelligence analysts is extended to assist analysts in the cyber defense domain. Secondly, a widely-used military targeting tool is leveraged to develop a methodology for prioritizing IDS/IPS alerts. To our knowledge, this is the first work to apply this targeting methodology in developing intrusion detection rulesets. Finally, a fuzzy rule set is designed that translates subject matter experts' (SME) knowledge into actionable knowledge [2] for cyber defenders.

The remainder of the paper is organized as follows: Sect. 2 provides an overview of cyber security in organizations and highlights an opportunity to improve business processes. Section 3 provides a brief discussion of fuzzy logic, related work in the cyber security domain, and similarities between military intelligence analysts and cyber defense analysts. Section 4 introduces the targeting methodology being applied in a unique manner to cyber defense. The Fuzzy Logic Utility Framework details are discussed in Sect. 5. Conclusions and future work are presented in Sect. 6.

2 Cyber Security in Organizations

Computers, information technology, and networking have had a tremendous impact on society and have transformed business, academia and military organizations. Our increasing reliance on computer information and communications networks dictates that systems must be secure, reliable and robust. Industry, financial institutions, academia, retailers and governments all have experienced security breaches of their enterprise networks [3–5]. Cyber security in organizations has become increasingly important in terms of protecting intellectual property, individuals' privacy, and business reputation.

Enterprise networks are often geographically dispersed, with potentially thousands of users, multiple types of network servers, and a wide array of networking equipment [6]. Considering the complexity and diversity of an enterprise network's resources and configurations, it is easy to see the cyber defender's challenge in developing and maintaining situation understanding of the network. It is unreasonable to expect the cyber defender to comprehend the enterprise network's complete operational picture and quickly ascertain the implications of sensor alerts.

Many organizations have CND or CSOC capabilities. These security centers are equipped with technologies and equipment that collect network data (flows and behaviors), perform correlation, and fire alerts when events of interest occur. Cyber defenders use these tools, threat intelligence, and their expert knowledge to gather contextual information to further evaluate the event(s). It is important to realize that no single cyber defender can be expected to have high proficiency in every technology used in the CND center or CSOC, or to know the "normal" behaviors of devices on every segment or within every enclave of the network he is responsible for monitoring.

An approach is needed that better supports the cyber defender by reducing cognitive load and decreasing the mean time to remediate; prioritizing sensor alerts will accomplish both objectives.

A recommended best practice for creating continuous improvement in CND functions and tools is to define a security operations center that encompasses essential elements of the CND. Reference [7] recommends placing all CSOC functions in a single CND organizational unit. Creating a tight coupling of all CND functions promotes efficiency, maximizes CND resources and enhances situational understanding and awareness for all cyber defenders and the Information Assurance Manager (IAM) staff.

The IAM staff is typically responsible for maintaining awareness of all network devices, configurations and enclaves, user privilege levels, and regulatory compliance. The IAM staff is also responsible for leading and maintaining contingency planning and continuity of operations planning, and is, therefore likely to be the entity most aware of senior leadership's intentions regarding mission critical operations. We view these as primary reasons for involving the IAM staff in the development of rules for prioritizing sensor alerts.

2.1 Process Improvement Strategy

Our approach will benefit organizations that lack the structuring recommendation in [7] by combining enterprise-wide (strategic) knowledge with real-time monitoring/incident coordination and response (operational) experts. This approach also applies to the computer network defense service provider (CNDSP) model, which is mandated by [8] as the foundation of the Department of Defense (DoD) Cyber Incident Handling Program.

This approach involves staff with expert strategic awareness and understanding of the enterprise network infrastructure and staff with expert knowledge of network monitoring technologies and threat intelligence. Further, it can be customized to fit any organization's desire or need to emphasize protection of particular assets, enclaves, or systems that support a specific mission.

The following list summarizes the rationale and steps for our approach.

1. Systems Security Plans (SSP) and Risk Assessment planning, contingency plans, etc., typically are the responsibility of the Information Assurance Manager (IAM) staff.
2. Cyber defenders stay abreast of cyber news, cyber intelligence, malware analysis, threat assessment, IDS and network monitoring tools.
3. The IAM staff use the FLUF application to define fuzzy rules for prioritizing alerts.
4. Alert Priorities via FLUF are intended to be reviewed on a weekly or bi-weekly basis. The ruleset must be kept current and synchronized with the organization's contingency plans and missions in order to be effective.

Our approach is designed to be compatible with any tool that offers a user-configurable rule-based engine. This approach can be used with either correlated or raw alerts.

3 Fuzzy Logic and Cyber Security

Human intelligence allows us to routinely make complex decisions when presented with incomplete, sometimes contradictory, and imprecise information. Artificial Intelligence (AI) techniques are used heavily in CND tools that are intended to help cyber defenders cope with the sheer amount of data [9, 10]. CND tools can unintentionally increase analyst workload by generating a high number of alerts that may be of little or no consequence but nevertheless must be investigated. Einstein 2 sensors, the IDS deployed by the Department of Homeland Security, "generate approximately 30,000 alerts on a typical day" and each alert is "evaluated by DHS security personnel to determine whether the alert represents a compromise" [11]. Reference [12] notes that some IDS trigger thousands of alerts each day, "99 % of which are false positives".

False positives represent imperfect information. The overwhelming amount of imperfect information makes fuzzy-based systems an appropriate choice for managing sensor alerts. Examples of imperfect information in the cyber domain include information that is imprecise, incomplete, and/or uncertain.

3.1 Fuzzy Logic's Appeal

As within many other domains, our need is to draw conclusions from imperfect information. We typically have to use words, or *linguistic variables*, to represent data and relationships. This suggests fuzzy logic would be an appropriate mechanism for us to use to develop the system for prioritizing sensor alerts.

In 1965, a method of computing qualitatively instead of quantitatively was introduced. Lotfi Zadeh formally defined multi-valued, or fuzzy, set theory [13]. Fuzzy sets provided a method for programming computers to draw deductive inferences.

Approximate reasoning can be defined as the "process or processes by which a possibly imprecise conclusion is deduced from a collection of imprecise statements" [14]. Inference systems to be used for approximate reasoning can easily be developed using fuzzy logic.

Using fuzzy sets, the approximate relationships between the input(s) (antecedent) and the output (consequent) are captured by the fuzzy rules of inference. A fuzzy rule with two antecedents and one consequent has the form "If X is A and Y is B then Z is C" where A and B are fuzzy sets over the two distinct input domains and C is a fuzzy set over some output domain [15].

The well-known advantages of fuzzy systems include their ability to handle imprecise, uncertain, and vague information; model complex non-linear systems which might be impossible to do mathematically; represent human decision making by handling vague data; and provide robustness due to being able to handle noisy and/or missing data [16]. These characteristics make a fuzzy-based approach highly attractive for our problem domain.

3.2 Fuzzy Logic and IDS—Related Work

The literature offers many examples of applying fuzzy logic to IDS. Reference [17] used fuzzy logic to reduce the number of correlated alerts based on alert occurrence frequency and the ratio of the attack to the total number of attacks. Related work using fuzzy logic focuses on reducing the number of false positives, with little or no effort applied toward prioritizing the remaining alerts [18].

Other closely related work was proposed by [19] in which fuzzy logic was used to score network asset valuation. Their proof-of-concept model indicates that the relative value of network assets can be accurately derived by combining knowledge of "experienced analysts" and those possessing expert knowledge of the monitored network. The author does not apply the scored network assets to any detection system.

Reference [20] used a fuzzy-based approach to prioritize alerts for signature-based IDS only. Their work does not address anomaly-based systems.

Work in [21] provides a comprehensive review of network event prioritization efforts and details their unique Event Prioritization Framework. Their framework offers a complete tool suite for network and host event correlation and prioritization. Data from industry and government standard sources are ingested, fused and processed to provide automated event prioritization. Due to the high level of automation of source data, network asset criticality in relation to specific missions cannot be determined within their framework.

Fuzzy logic has been used to classify alert severity as high, medium and low [22, 23]. It is easy to see this can result in having a large number of "high" severity alerts and the problem of prioritizing the high risk alerts remains.

Our approach to prioritizing alerts distinguishes itself from the works reviewed above by (1) considering the criticality of network assets in relation to the mission, (2) requiring no changes to existing tool configurations or data sources, and (3) integrating with behavior, anomaly, or signature-based alerts. Another distinguishing feature of our approach is the use of the CARVER military targeting methodology; the methodology will be explained in Sect. 4.

3.3 Intelligence Analysts and Cyber Analysts

As mentioned above, a primary obstacle faced by cyber defense analysts is handling imprecise, uncertain, and vague information within a time critical environment. An analogous situation occurs within the domain of military intelligence. In our context, intelligence is defined as the "product resulting from the collection, processing integration, evaluation, analysis, and interpretation of available information" [24]. In the domains of both the cyber analyst and military intelligence analyst, one can see that *analysis* and *interpretation* are processes which largely fall on the human while automation, if available, only assists with the other listed processes.

Another similarity of the cyber defender and the military intelligence analyst is the consequential results their decisions have on operations and people beyond their immediate circle. Further, there is a shortage of qualified and experienced intelligence analysts, just as the demand for cyber defenders has outgrown the supply. In both cases, these shortages impact our national security [25]. High stakes decision making under uncertainty and time constraints, and access to tremendous amounts of information are characteristic of the environments in which cyber defenders and intelligence analysts operate.

By this point it should be clear that the problems within the decision cycle for cyber analysts are very similar to those of military intelligence analysts. Having recognized that, our work seeks to leverage recent research done with respect to developing a fuzzy-based automated decision aid for military intelligence analysts. The system is a tool for supplying intelligence analysts with relevant, reliable information, rank ordered by the information's importance within a specific mission context. Recent validation has shown that this *Value of Information* (VoI) metric positively affects decision making quality and workload in military intelligence analyst teams at the tactical level [26–28].

The fuzzy VoI system has been described in detail in previous publications [29–33]; space constraints do not allow the specifics to be repeated here. Basically, the system is composed of a two-level Fuzzy Associative Memory (FAM) architecture where each of the two FAMs has two inputs and one output. Thus, each FAM is a two-dimensional table (matrix) where each dimension corresponds to one of the input domains. Fuzzy if-then rules are represented in the FAM by allowing each cell to be indexed by the fuzzy sets that comprise the specific input domains (the fuzzy rule antecedents); the contents of the cell then represent the associated output based on these inputs (the fuzzy rule consequent). The fuzzy rules were developed through a combination of regulatory guidance and a significant knowledge elicitation process with subject matter experts [34]. The efficacy of the system has been validated by both SME feedback and human performance model simulation results [26, 27].

4 CARVER Targeting Methodology

This section discusses the CARVER targeting methodology and provides a foundation for understanding how it, combined with the fuzzy logic construct, offers a mechanism to rate alerts based on factors specific to a particular networking environment and its mission/business function.

As discussed in Sect. 1, the demand and critical need for cyber defenders outpaces a ready supply. Our reliance on technology and networks has caused US policy makers to view cyberspace as a domain in military operations [35]. These two conditions suggested researching target selection approaches used by small units or teams.

A target analysis process developed by the US Army's Special Operations Forces (SOF) emerged as good choice. It has historically been used as an offensive targeting tool to prioritize targets within complex systems. Since the 9/11 attacks and the rise in concern of terrorism in the US, it has been adapted for use in vulnerability and risk assessments, primarily in the food industry [36].

CARVER stands for Criticality, Accessibility, Recuperability, Vulnerability, Effect, and Recognizability; it is used to determine the military value and priority of potential targets. Each of the six components is rated and placed in a matrix to rank systems or subsystems that are considered eligible targets. The elements that rank at or above a designated threshold are deemed suitable for attack.

The following definitions of each component selected for this initial study are taken from [37]:

- Criticality reflects the degree to which the target's "destruction, denial, disruption" and damage will impair the adversary.
- Accessibility is an estimate of the SOF to physically or indirectly reach the target. SOF must also be able to clear out of the target area without detection.
- Recuperability supports the Criticality element. If a target can be easily repaired or replaced, it may garner a low rating.
- Vulnerability reflects the degree to which SOF is able to inflict the desired level and type of damage.
- Effect is a rating of the impact the target destruction, denial, or disruption will have on the adversary.

- Recognizability reflects the level to which the target can be distinguished from similar objects in the area.

CARVER is a semi-quantitative risk assessment. According to [38], this method of assessment is “most useful in providing a structured way to rank risks according to their probability, impact or both (severity)”.

As previously explained, the fuzzy logic VoI construct was developed to assist intelligence analysts with identifying valuable information under time and resource constraints. We believe that by selecting the appropriate CARVER components, the advantages of the fuzzy-based VoI architecture can be directly translated to the cyber domain to provide similar assistance to cyber analysts.

5 Fuzzy Logic Utility Framework

As discussed previously, the domains of the cyber defender and military intelligence analyst are similar in terms of having to make decisions under uncertainty with incomplete and/or possibly inaccurate information. It is with this understanding that our proposed Fuzzy Logic Utility Framework (FLUF) has been mirrored after the fuzzy-based VoI system developed for military intelligence analysts.

The current state of FLUF is that of a prototype, proof-of-concept decision support tool to assist cyber intelligence analysts in addressing intrusion detection alerts based on the order of their severity. The FLUF is used to determine the Alert Priority Rating (APR), which arranges IDS/IPS alerts in high to low order of severity. The FLUF considers the criticality of the affected asset, the level of access needed to infiltrate the asset, and the effect of the affected asset’s compromise to the enterprise or mission.

Due to the effectiveness of the VoI system, the small computational impact, and rule set economy [28], the same fuzzy logic architecture was adopted for the FLUF system. Extending the foundational structure of the VoI system to other domains is expected to provide support to decision makers in terms of reduced cognitive load, increased time in performing analysis rather than foraging for information [27] and potentially reduce the number of sensor alerts. Further, it provides an excellent opportunity to exercise the software engineering principle of reuse.

The starting point for developing any fuzzy rule-based system is to decompose the input and output domains into fuzzy sets. As an example, the decomposition of the FLUF *Criticality* domain is shown in Fig. 1. It is decomposed into five overlapping fuzzy sets, with each fuzzy set representing a classification. An element in the domain has some grade of membership, from 0 to 1 inclusive as shown on the y axis, in each fuzzy set in the domain. The membership function determines the grade of membership; the shape of the fuzzy sets determines the membership function.

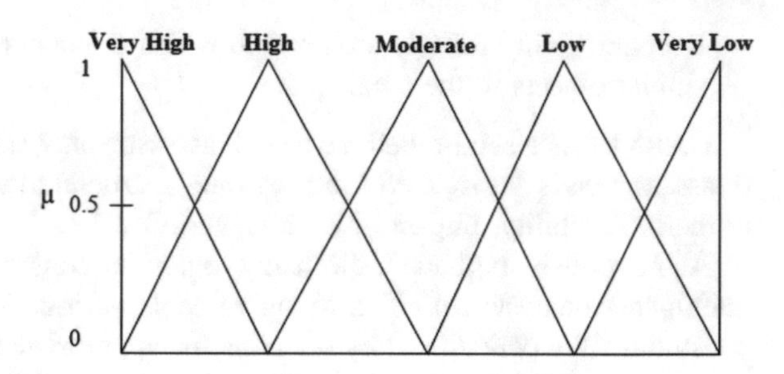

Fig. 1 Criticality domain

In Fig. 1, any input within the domain will belong to at most two fuzzy sets; that is, any input will have non-zero membership in no more than two fuzzy sets. This means that, for each input, the antecedents for at most two fuzzy rules associated with that domain will be satisfied.

Further, the sum for all membership values in the sets to which any input belongs will equal 1. The decomposition shown in Fig. 1 illustrates the membership functions as isosceles triangles with bases of the same width, and has been called a TPE system (triangular decomposition with evenly spaced midpoints) [39].

Note that the shape of the fuzzy sets does not have to be triangular as shown in Fig. 1. Also, the fuzzy sets decomposing a domain do not have to overlap in a regular pattern, nor does the sum have to be 1 for membership values for all sets in which an element is a member.

5.1 Designing the Fuzzy Rule Base

Requirements engineering and user-centered design practices recommend engaging the proposed users early in the system design in order to better understand their needs [40]. Following the iterative user-centered design process described in [41], we derived understanding of our users' environments, processes and tasks in the *early envisioning* phase. In this research, our users are the SMEs from the IAM and CND organizations.

The majority of the SMEs were very familiar with the DoD Risk Assessment Guide [38], and the entire SME pool was well acquainted with the Common Vulnerability Scoring System (CVSS), its exploitability metrics, and vulnerability impact scope. We immediately recognized similarities between the CVSS and CARVER scales, and chose to nominate the *criticality* and *accessibility* CARVER components as two of the FLUF input domains. These were presented to the SMEs in a follow-up discussion. The SMEs agreed with those choices and collaboratively selected *effect* as the third input domain for the initial study.

The three domains were presented to the SMEs as follows:

Criticality—Can the organization complete its mission if the asset fails to meet the basic security principles of Confidentiality, Integrity and Authentication?

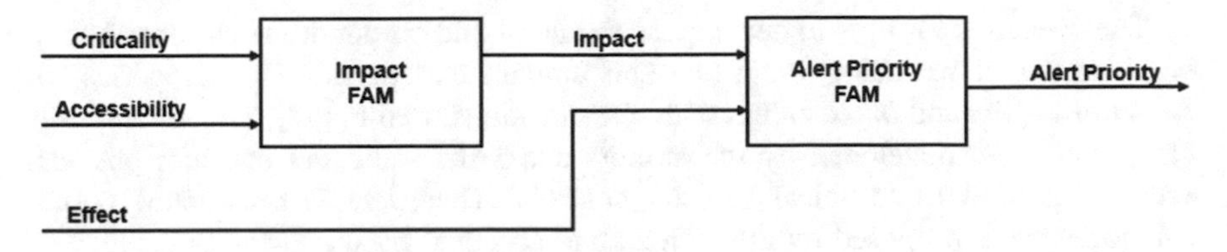

Fig. 2 FLUF System Architecture

Accessibility—How easily can the asset be reached? How secure is its configuration and are other technical and/or physical defenses employed to protect it? Does access require insider knowledge and physical presence?

Effect—An initial effect of a data breach may be the company's reputation. If intellectual property was compromised, the effect may be even greater in the long term.

The design decision to combine the *criticality* and *accessibility* domains to produce an *impact* determination was influenced by CVSS' exploitability metrics. The SMEs agreed that the essence of those metrics, Attack Vector, Attack Complexity, Privileges Required, and User Interaction, were reflected in the descriptions provided for the *criticality* and *accessibility* domains.

These two domains provide the inputs to the first of two Fuzzy Associative Memories (FAMs) shown in Fig. 2.

The SMEs agreed that combining the *effect* input with the *Impact* FAM output would provide context that is much needed to understand the value of a network asset. The output of the *Impact* FAM, and the *effect* measure are then input to the *Alert Priority* FAM, which in turn produces the Alert Priority Rating (APR). The overall output of the entire system, the APR, would then be included in an IDS rule alert, Snort for example, that when triggered, would indicate to CND analysts which devices should receive attention first.

At this point, the input and output domains and the overall architecture for the FLUF were decided; the next step was to decompose the input and output domains into fuzzy sets, thereby determining the language of the rule base. That is, the decompositions define the possible antecedents and consequents for the fuzzy rules.

The linguistic decomposition of the *criticality* domain is as shown in Fig. 1. The *accessibility* domain is similarly decomposed into five fuzzy sets, with linguistic terms that range from "easily accessible" to "not accessible". The *effect* input domain is decomposed into three fuzzy sets: High (extensive), Medium (wide-ranging) and Low (limited, still functional). The *Impact* FAM output domain is decomposed into nine fuzzy sets that linguistically range from "very low" to "very high" impact. There are nine output fuzzy sets for the *Alert Priority* FAM that range from "negligible" to "urgent". Note that all fuzzy sets use triangular membership functions that follow the TPE requirements mentioned previously. This decision was made to leverage the advantages of the previously developed fuzzy-based VoI system, which include computational efficiency and facilitation of the knowledge acquisition process with the SMEs [28].

The decisions for how to decompose the input and output domains came from a combination of regulatory guidance and domain experience. The categories for *criticality, effect* and *impact* align with classifications used in [42]. The *accessibility* categories were developed by the authors and IAM staff, and consider physical security as well as technical security controls. The *Alert Priority* FAM output categories were proposed by the authors and accepted by the SMEs.

In addition to reusing the favorably received VoI architecture, the resulting FLUF system architecture provides significant flexibility in changing CARVER components or introducing Risk Management Framework (RMF) components. Since the domains were decomposed to reflect the assessment scales in [38], any of the RMF risk elements can be substituted in the FLUF.

5.2 Fuzzy Rule Base Review

After the FLUF architecture was defined, the SMEs were provided a 5 × 5 square matrix with *criticality* categories as the rows and *accessibility* categories as the columns. They received the *Impact* FAM scale, with categories ranging from "very high" through "very low".

The SMEs were provided a network diagram representative of a small business enterprise network. They were asked to rank each device's *criticality* and *accessibility* using their knowledge of best security practices and regulations governing network architecture, device configurations and physical security. From this interaction, the authors wrote the *Impact* FAM rules and distributed them to the SMEs for comment. The SMEs analyzed and accepted the rules with only minor changes.

Based on having the *criticality* and *accessibility* domains each decomposed into five fuzzy sets, the resulting *Impact* FAM consists of 25 rules. Example actual rules include:

- If *Criticality* is Very High, and *Accessibility* is Easily Accessible, then *Impact* is Very High.
- If *Criticality* is Very High, and *Accessibility* is Accessible, then *Impact* is High.
- If *Criticality* is Moderate, and *Accessibility* is Easily Accessible, then *Impact* is High to Moderate.

The SMEs then rendered their opinions of *effect* to the mission/business function. The values for *effect* range from High (1) to Low (3), and are combined with the *impact* input to derive the APR result. The linguistic decompositions for *effect* and the descriptions were provided by the authors and accepted by the SMEs.

After some initial discussion with the SMEs, the rules for the *Alert Priority* FAM were developed based on the *effect* and *impact* ratings previously produced by the SMEs. These rules were distributed for comment; no changes were requested.

Since the impact domain has nine fuzzy sets while the effect domain has three fuzzy sets, the Alert Priority FAM consists of 27 fuzzy rules. Example rules in this FAM include:

- If *Effect* is High, and *Impact* is Very High, then *Alert Priority* is Critical to Urgent.
- If *Effect* is High, and *Impact* is High, then *Alert Priority* is Critical.
- If *Effect* is Medium, and *Impact* is High, then *Alert Priority* is High to Critical.

At this point the SMEs were satisfied that the resulting fuzzy rules in both FAMs accurately and appropriately reflected the reasoning methodology for correctly prioritizing alerts. Note that space and other limitations preclude the exhaustive listing of all fuzzy rules within the FLUF architecture. It should also be noted that we do not claim that the rule bases have been "verified" by the SME review as described above. However, given that the FLUF is currently a prototype, proof-of-concept system, we consider the system to be "validated" by the user. Future experiments are certainly planned to more exhaustively and scientifically validate, as well as verify, the system.

5.3 Additional Remarks Regarding FLUF and APR

To further ensure that the users (SMEs) were satisfied that the resulting system would meet their needs, survey questions were distributed and group discussions were held. The SMEs were asked if they believed the proposed process improvement and Alert Priority Rating would have a positive effect on CND responses to events and incidents. 100 % of the SMEs agreed this research would improve response time to incidents threatening critical assets and improve CND analysts' situational awareness of the networks they monitor since the IAM staff provided assessments of which assets were most critical.

One of the SMEs, a system administrator, observed that without understanding the layout and structure of the network they are defending, and the critical assets, they are "blindly" analyzing traffic. The FLUF architecture provides, and indeed is designed to require, the tuning of IDS sensor rule sets for particular subnets or devices, depending on asset criticality.

Another SME noted that a compromise on a desktop machine of the CEO could be of greater concern to an organization than an alert firing on a piece of key infrastructure. The proposed process improvement and the Alert Priority Rating provide means of addressing that situation.

In addition to discussions with the SMEs, a very rudimentary and preliminary empirical experiment was performed. Alert priority ratings, modeled after those that would be produced by a particular FLUF implementation, were manually plugged into some Snort rules and applied to part of the VAST 2011 Challenge data set. The results showed that of the 49,000 alerts occurring over an 8-h period, only 3 of them were truly high priority. It is also interesting to consider (and study in the future) how many of those alerts would have been negated if any one of the 3 serious alerts had been attended to first.

These remarks and results reinforce our belief that the FLUF and APR will support CND analysts in their overwhelming task of addressing the most important alerts first, and provide the necessary proof-of-concept confirmation to justify continued investigation and evolution of the FLUF system.

6 Conclusions and Future Work

The FLUF approach is entirely system-agnostic; it will integrate with signature, anomaly or behavior-based IDS methods. We believe that relying on expert human judgments regarding network asset criticality is an essential element of prioritizing sensor alerts and accordingly increasing defense of those assets. We expect the FLUF approach to prioritizing alerts to increase defenses of critical network assets and decrease the mean time to remediate actual network incidents.

The work presented in this paper offers several main contributions. First, a fuzzy logic architecture previously constructed to assist military intelligence analysts is extended to assist analysts in the cyber defense domain. Secondly, a widely-used military targeting tool is leveraged to develop a methodology for prioritizing IDS/IPS alerts. To our knowledge, this is the first work to apply this targeting methodology in developing intrusion detection rulesets. Finally, a fuzzy rule set is designed that translates subject matter experts' (SME) knowledge into actionable knowledge for cyber defenders.

In summary, benefits of the FLUF include:

- Valuation of network assets as an element of the priority rating; informing contingency plans and supporting synchronization with operation.
- FLUF rules can be changed as dictated by mission, changes in the operational environment, and discovery of new knowledge (infiltrations, infections, attack campaigns) evolves.
- FLUF rules are synchronized with changes in network and asset configurations, and mission functions.
- Increased collaboration among the IAM/CND/CSOC teams to spread information assurance/network security expertise throughout the organization.
- Expansion of cyber defenders' domain expertise.

Future work will focus on executing an experiment on the VAST 2011 Challenge data set to collect empirical evidence demonstrating the effects of the APR. The VAST 2011 solution guide details attacks against the network described in the Challenge. Snort rules designed for use in the VAST Challenge will be augmented with the APR to quantify the effect and efficacy of the FLUF and APR on that data set.

Additional research could include studying how the FLUF and APR affect the number of subsequent alerts, and investigating new feature-value pairs in the *criticality* fuzzy domain in relation to mission priorities. Each of these activities could contribute to information assurance research efforts within the organization.

References

1. Pegna, D. L. (2015). *Big data sends cybersecurity back to the future*. Retrieved June 15, 2015, from http://www.computerworld.com/article/2893656/the-future-of-cybersecurity-big-data-and-data-science.html.
2. Leedom, D. K. (2004). *Analytic representation of sensemaking and kowledge management within a military C2 organization*. Vienna, VA: Evidence Based Research Inc.
3. Ramanan, S. (2015). Top ten security breaches of 2015. *Forbes.com*. Retrieved from http://www.forbes.com/sites/quora/2015/12/31/the-top-10-security-breaches-of-2015/#2f01eff01f76.
4. (ISC)^2 US Government Adviory Council Bureau. (2015). *There were so many data breaches in 2015. Did we learn anything from them?* Retrieved December 29, 2015, from http://www.nextgov.com/technology-news/tech-insider/2015/12/there-were-so-many-data-breaches-2015-did-we-learn-anything-them/124780/.
5. Cirilli, K. (2014). *Home depot breach costs doubled target's*. Retrieved October 30, 2014, from http://thehill.com/policy/finance/222340-home-depot-breach-costs-doubled-targets.
6. Juniper Networks. (2014). *Network configuration example midsize enterprise campus solution*.
7. Zimmerman, C. (2014). Ten strategies of a world-class cybersecurity operations center. In M. A. Bedford (Ed.), *MITRE corporate communications and public affairs*. Appendices.
8. Joint Staff. (2012). *CJCSM 6510.01B department of defense cyber incident handling program*.
9. Lee, D., Hamilton, S. N., & Hamilton, W. L. (2011). Modeling cyber knowledge uncertainty. In *2011 IEEE Symposium on Computational Intelligence in Cyber Security (CICS)*.
10. Alrajeh, N. A., & Lloret, J. (2013). Intrusion detection systems based on artificial intelligence techniques in wireless sensor networks. *International Journal of Distributed Sensor Networks, 2013*, 6.
11. Anonymous. (2015). Department of homeland security, *Einstein*. Retrieved January 15, 2016, from http://www.dhs.gov/einstein.
12. Julisch, K., & Dacier, M. (2002). Mining intrusion detection alarms for actionable knowledge. In *Proceedings of the Eighth ACM SIGKDD International Conference on Knowledge Discovery and Data Mining* (pp. 366–375).
13. Zadeh, L. A. (1965). Fuzzy sets. *Information and Control, 8*, 15.
14. Zadeh, L. A. (1973). Outline of a new approach to the analysis of complex systems and decision processes. *IEEE Transactions on Systems, Man and Cybernetics,* 28–44.
15. Hammell II, R. J., Powell, J., Wood, J., & Christensen, M. (2010). Computational intelligence for information technology project management. In *Intelligent Systems in Operations: Methods, Models and Applications in the Supply Chain* (p. 80).
16. Yen, J., & Langari, R. (1998). *Fuzzy logic: Intelligence, control, and information*. Prentice-Hall, Inc.
17. Huang, C., Hu, K., Cheng, H., Chang, T., Luo, Y., & Lien, Y. (2012). Application of type-2 fuzzy logic to rule-based intrusion alert correlation detection. *International Journal Innov Computing Inform and Control, 8*, 65–74.
18. Alshammari, R., Sonamthiang, S., Teimouri, M., & Riordan, D. (2007) Using neuro-fuzzy approach to reduce false positive alerts. In *Fifth Annual Conference on Communication Networks and Services Research. CNSR'07.* (pp. 345–349)
19. Leung, H. (2015). An asset valuation approach using fuzzy logic. In *SPIE Sensing Technology + Applications*.
20. Alsubhi, K., Al-Shaer, E., & Boutaba, R. (2008). Alert prioritization in intrusion detection systems. In *IEEE Network Operations and Management Symposium. NOMS 2008* (pp. 33–40).
21. Kim, A., Kang, M., Luo, J. Z., & Velasquez, A. (2014). A framework for event prioritization in cyber network defense. DTIC Document.
22. Tabia, K., Benferhat, S., Leray, P., & Mé, L. (2011). Alert correlation in intrusion detection: Combining ai-based approaches for exploiting security operators' knowledge and preferences. In *Security and Artificial Intelligence (SecArt)*.

23. Alsubhi, K., Aib, I., & Boutaba, R. (2012). FuzMet: A fuzzy-logic based alert prioritization engine for intrusion detection systems. *International Journal of Network Management, 22*, 263–284.
24. Joint Staff. (2015). *Joint Publication 1-02 Department of Defense Dictionary of Military and Associated Terms.*
25. Libicki, M. (2014). Shortage of cybersecurity professionals poses risk to national security. *Rand.org*. Retrieved from http://www.rand.org/news/press/2014/06/18.html.
26. Newcomb, E. A., & Hammell II, R. J. (2013). A method to assess a fuzzy-based mechanism to improve military decision support. In *14th ACIS International Conference on Software Engineering, Artificial Intelligence, Networking and Parallel/Distributed Computing (SNPD)* (pp. 143–148).
27. Hanratty, T. P., Newcomb, E. A., Hammell II, R. J., Richardson, J. T., & Mittrick, M. R. (2016). A fuzzy-based approach to support decision making in complex military environments. *International Journal of Intelligent Information Technologies (IJIIT), 12*, 1–30.
28. Newcomb, E. A., & Hammell II, R. J. (2012). Examining the effects of the value of information on intelligence analyst performance. In *Proceedings of the Conference on Information Systems Applied Research ISSN* (p. 1508).
29. Hammell, R. J., Hanratty, T., & Heilman, E. (2012). Capturing the value of information in complex military environments: A fuzzy-based approach. In *2012 IEEE International Conference on Fuzzy Systems (FUZZ-IEEE)* (pp. 1–7).
30. Hanratty, T. P., Hammell II, J.R., Bodt, B.A., Heilman, E.G., & Dumer, J.C. (2013). Enhancing battlefield situational awareness through fuzzy-based value of information. In *2013 46th Hawaii International Conference on System Sciences (HICSS)* (pp. 1402–1411).
31. Hanratty, T. P., Dumer, J. C., Hammell II, R. J., Miao, S., & Tang, Z. (2014). Tuning fuzzy membership functions to improve value of information calculations. In *2014 IEEE Conference on Norbert Wiener in the 21st Century (21CW)* (pp. 1–7).
32. Miao, S., Hammell II, R. J., Hanratty, T., & Tang, Z. (2014). Comparison of fuzzy membership functions for value of information determination. In *MAICS* (pp. 53–60).
33. Miao, S., Hammell II, R.J., Tang, Z., Hannratty, T. P., Dumer, J. C., & Richardson, J. (2015). Integrating complementary/contradictory information into fuzzy-based voi determinations. In *2015 IEEE Symposium on Computational Intelligence for Security and Defense Applications (CISDA)* (pp. 1–7).
34. Hanratty, T., Heilman, E., Dumer, J., & Hammell II, R. J. (2012). Knowledge Elicitation to Prototype the Value of Information. In *Midwest Artificial Intelligence and Cognitive Science Conference* (p. 173).
35. Joint Staff. (2014). Joint Publication 3-13 Information Operations.
36. Catlin, M., & Kautter, D. (2007). An overview of the Carver Plus Shock method for food sector vulnerability assessments. *USFDA, editor. USFDA.* (pp. 1–14).
37. U. S. Army. (2012). *Army Doctrinal Reference Publication (ADRP) 3-05*, Special Operations ed. Washington, DC: Headquarters, Department of the Army.
38. Microbiological Risk Assessment Series. (2009). No. 17, Chapter 4. Semi-quantitative risk characterization. ISBN 978 92 4 154789 5.
39. Sudkamp, T., & Hammell, R. J, I. I. (1994). Interpolation, completion, and learning fuzzy rules. *IEEE Transactions on Systems, Man and Cybernetics, 24*, 332–342.
40. Sommerville, I., & Kotonya, G. (1998) *Requirements engineering: Processes and techniques.* Wiley.
41. Wassink, I., Kulyk, O., van Dijk, B., van der Veer, G., & van der Vet, P. (2009). Applying a user-centered approach to interactive visualisation design. In *Trends in Interactive Visualization* (pp. 175–199). Springer.
42. NIST Computer Security Division. (2010). *Guide for applying the risk management framework to federal information systems* (Vol. 800-37 rev1). NIST Special Publication.

A Framework for Requirements Knowledge Acquisition Using UML and Conceptual Graphs

Bingyang Wei and Harry S. Delugach

Abstract UML provides different models for understanding and describing the requirements of a system. The completeness of each model with respect to other models is critical to further analysis of the requirements and design. One problem that always plagues modelers is the acquisition of requirements knowledge for building models. In this paper, we present a knowledge-based framework to drive the process of acquiring requirements for each UML model. This framework is based on a central knowledge representation, the conceptual graphs. A set of partially complete UML models is first converted to conceptual graphs to form a requirements knowledge reservoir; then this knowledge reservoir is used to generate each UML model by transforming conceptual graphs back to UML notations. This bidirectional transforming process enables the discovery of additional requirements and possible missing requirements so that eliciting more requirements knowledge from modelers is made possible.

1 Introduction

Software requirements modelers build different types of models for a system under development. Each of them holds partial requirements from a particular view and all of them together constitute the overall description of the system. UML makes this multiple-viewed modeling technique possible by providing different types of diagrams. An important concern of modelers during multiple-viewed modeling is the acquisition of enough useful requirements to make a model complete. However,

B. Wei (✉)
Department of Computer Science, Midwestern State University,
Wichita Falls, TX 76308, USA
e-mail: bingyang.wei@mwsu.edu

H.S. Delugach
Department of Computer Science, University of Alabama in Huntsville,
Huntsville, AL 35899, USA
e-mail: delugach@cs.uah.edu

R. Lee (ed.), *Software Engineering Research, Management and Applications*, Studies in Computational Intelligence 654,
DOI 10.1007/978-3-319-33903-0_4

it is difficult for a modeler to know whether a model is complete or what requirements are missing in the current model [6]. Modelers need to be made aware of the missing requirements.

Delugach proposed the idea of conceptual feedback [5] which can provide prompts for the missing requirements of a model to modelers (Fig. 1). This approach is based on the requirements knowledge overlap among different models of the same system. During conceptual feedback, requirements in $Model_0$ to $Model_n$ in Fig. 1b are transformed to generate requirements needed for constructing a target model $Model_x$. This process introduces new requirements to $Model_x$ that a modeler is not aware of before the generation process, making it more complete. More importantly, such newly generated requirements may be used as stimulations to elicit more requirements about the model from modelers (This is shown as a backtracking from Specification to Elicitation in Fig. 1a). After eliciting and adding more new requirements in the model, this augmented model ($Model_x$) would in turn affect other models (dotted arrows in Fig. 1b), causing further generation and completion processes in other models. The process may repeat until no more new requirements knowledge can be acquired by transforming models, i.e., the set of models is internally complete and self-consistent.

In this paper, we present a knowledge-based framework to facilitate requirements acquisition for a set of UML models using conceptual feedback. This framework is based on a central knowledge representation, the conceptual graphs (CGs) [11]. CGs are based on existential conjunctive first-order logic. Since the models of a software system can be regarded as a collection of statements that evaluate to truth, i.e. assertional knowledge, CGs have been widely used to represent requirements

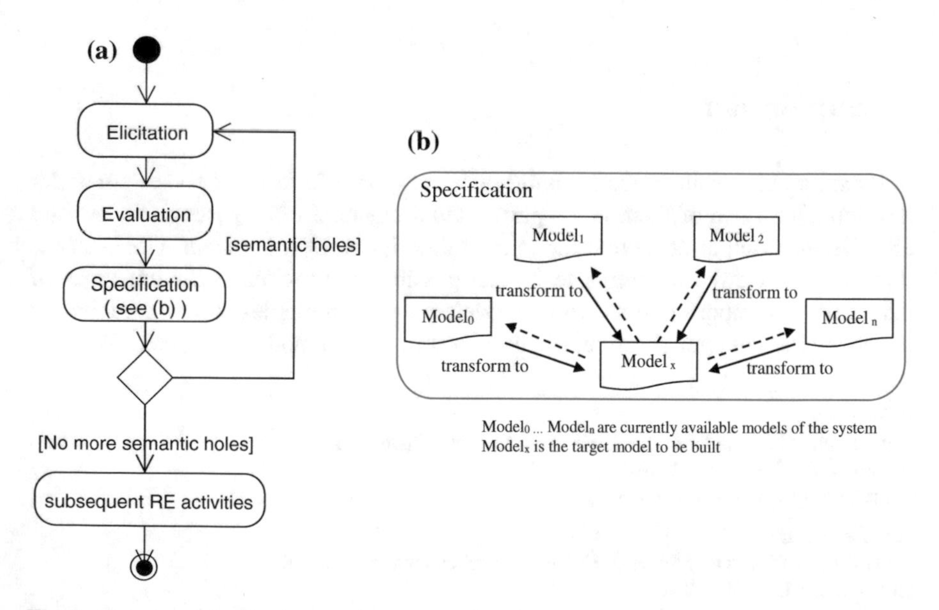

Fig. 1 Conceptual feedback in requirements engineering process

knowledge in UML diagrams [2, 5, 7, 12]. Two basic elements of CGs are concepts and relations. Concepts are represented by rectangles and relations are represented by ovals.

The rest of the paper is structured as follows. Section 2 provides an overview of the CGs-based knowledge acquisition framework and its key component. The bidirectional transformation between UML diagrams and CGs are described in detail in Sect. 3 using a case study. In Sect. 4, we evaluate the framework. Several issues, limitations, related work and future work are discussed in Sect. 5. In Sect. 6, we conclude the paper.

2 The CGs-Based Requirements Acquisition Framework

Our framework can be illustrated in Fig. 2. In this work, we include three types of UML diagrams in the framework. The framework consists of a CGs Reservoir where requirements knowledge of UML diagrams is stored in the form of CGs and a CGs Support which guides the bidirectional transformation between UML diagrams and CGs.

The process of generating UML diagrams with requirements acquisition opportunities consists of two major phases. In phase 1, a set of already developed but incomplete UML diagrams is converted to CGs in order to populate the CGs Reservoir (inward arrows in Fig. 2); in phase 2, each UML diagram is generated from the CGs Reservoir (the outgoing arrows). The generated UML diagrams during phase 2 keep their original requirements before phase 1 and are augmented with new requirements inferred from other UML diagrams in the set. A requirements knowledge acquisition process then starts in which modeler of each diagram provides necessary requirements to accommodate the newly generated requirements in the diagram, thereby adding more requirements.

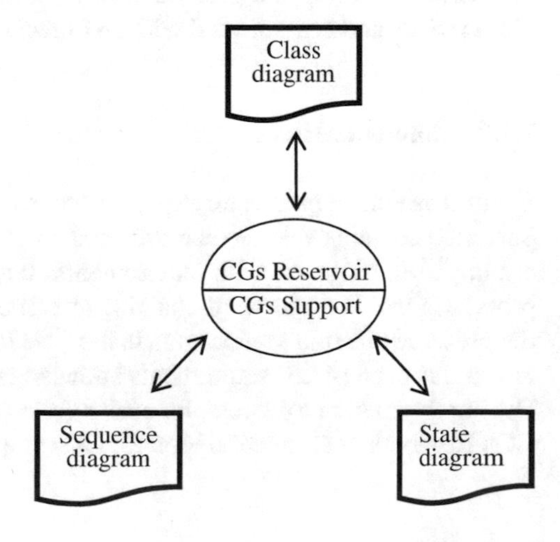

Fig. 2 CGs-based requirements acquisition framework

2.1 *The CGs Support*

A key component of our framework is the CGs Support. It defines, in CGs form, canonical graphs which are used to express the three types of UML diagrams in CGs, and inference rules for generating UML diagrams from the CGs Reservoir.

2.1.1 Canonical Graphs

In the framework, each type of UML diagram has a corresponding set of canonical graphs that describes its semantics in CGs. Canonical graphs are CGs used as templates to represent meaningful relationships among concepts in a particular type of UML diagram. The canonical graphs are not models themselves. UML diagrams are converted to CGs by instantiating corresponding sets of canonical graphs so that all of the semantics captured by the UML diagram are preserved in the central CGs Reservoir.

Canonical graphs of class diagrams are shown (Fig. 3). Interested readers can find canonical graphs of state diagrams and sequence diagrams in [16].

The canonical graph for a class is shown in Fig. 3a. Since a class describes a set of similar objects, the concept *ClassName: @forall* means "for all objects of this class." In this concept, *ClassName* denotes the name of a class and will be replaced with the name of a real class when this canonical graph is instantiated. An attribute is represented by a *T* type concept, which is related to the class concept through an *attribute* relation, while an operation is represented by an *Activity* type concept, which is related to the class concept by an *operation* relation. The *association* relates each object of this class to other objects. In Fig. 3a, only one attribute, one operation, and one association are shown in the canonical graph. Canonical graphs (b) and (c) of Fig. 3 represent composition and generalization, respectively.

The canonical graph in Fig. 3a represents "For each object of a class *ClassName* (*ClassName* is used as a placeholder), it has an attribute of type *T*, an operation of type *Activity* and is associated with an object of type *Object*."

2.1.2 Inference Rules

Besides canonical graphs for expressing meanings of UML diagrams, the CGs Support also contains rules which are used to infer requirements knowledge for generating UML diagrams with requirements acquisition opportunities. The generation process is based on a forward-chaining inference method. As in any logical inference, the presence of a rule's antecedent in the CGs Reservoir implies its consequent which represents the desired requirements knowledge used to build a target UML diagram. During the inference process, for each inference rule of the target UML diagram, the CGs Reservoir is scanned to look for CGs snippets that match the antecedent of the

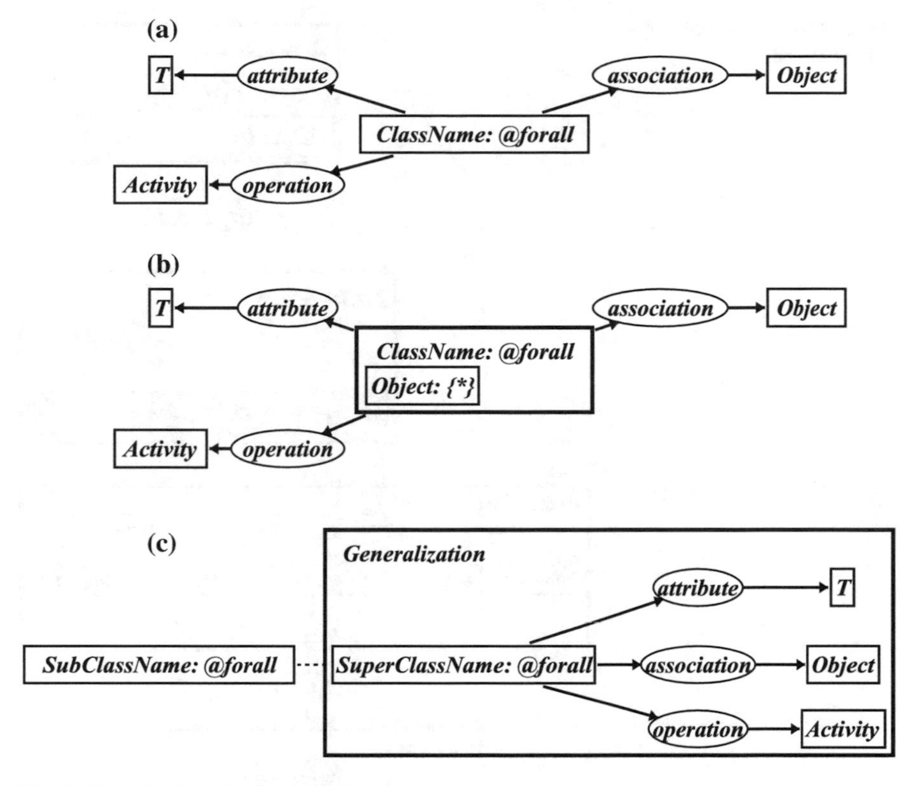

Fig. 3 Canonical graphs for class diagrams

rule. If a match is found, the consequent of the rule is asserted, thereby resulting in the derivation of requirements knowledge needed for building the diagram.

Two of the inference rules for class diagrams are shown in Figs. 4 and 5. When applying the Association Rule, the CGs snippet we are looking for in the CGs Reservoir is "an *Object* type concept receives a *Message* type concept issued by another *Object* type concept,'' which, if found, would imply the existence of an association between the two objects' classes. Note that the inferred CGs are colored in light gray in Fig. 5. Colors do not mean anything in CGs, they are used as a convenience here. Part of the inference process can be supported by the inference engine provided by CoGUI [1].

3 Case Study

In this section, we apply the requirements knowledge acquisition framework to a safety critical system, the Mine Safety Control System (MineSys) from [15]. In this system, sensors constantly collect environmental data of a mine and if a hazardous

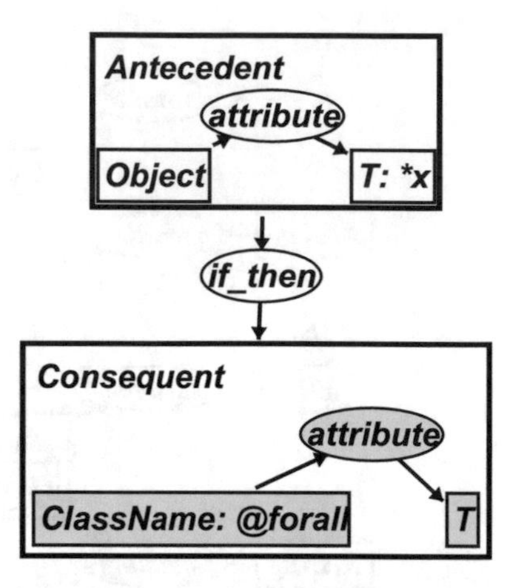

Fig. 4 Class inference rule 1: attribute rule

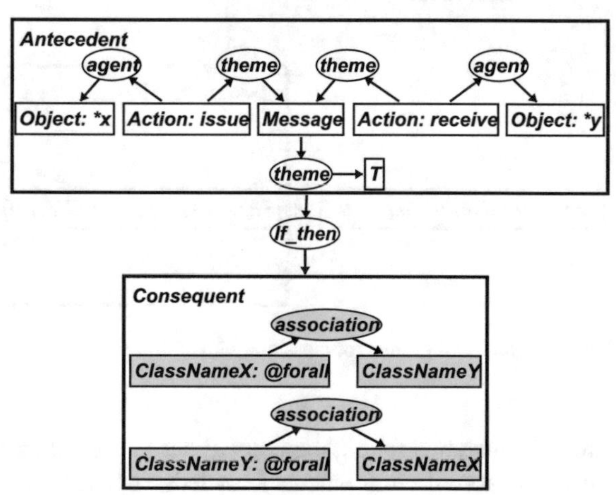

Fig. 5 Class inference rule 2: association rule

situation is detected, the system should enable the corresponding safety measures to protect miners. For example, a distributed real-time safety control system monitors water-level sensors to detect when the water in a sump is above a high or below a low level; pump should be turned on whenever the water has reached the high water level; it also monitors sensors detecting various gases in a mine, and alarm should be raised if any of these levels has reached a critical threshold. Based on the requirements, three initial UML models are built (see Figs. 6, 7 and 8). These diagrams have been significantly simplified for this paper.

In the framework, the three UML diagrams are converted to CGs according to their corresponding canonical graphs. Because of space limitations, only four classes in the class diagram are converted and shown in Fig. 9; CG of one transition in the state diagram is shown in Fig. 10; the sequence diagram's CG is shown in Fig. 11.

With the CGs Reservoir populated, the requirements acquisition process is able to start. For example, the modeler of the class diagram of MineSys is trying to make

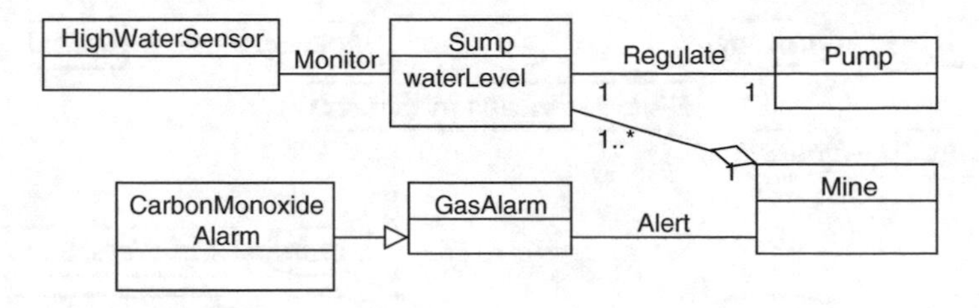

Fig. 6 Class diagram of MineSys

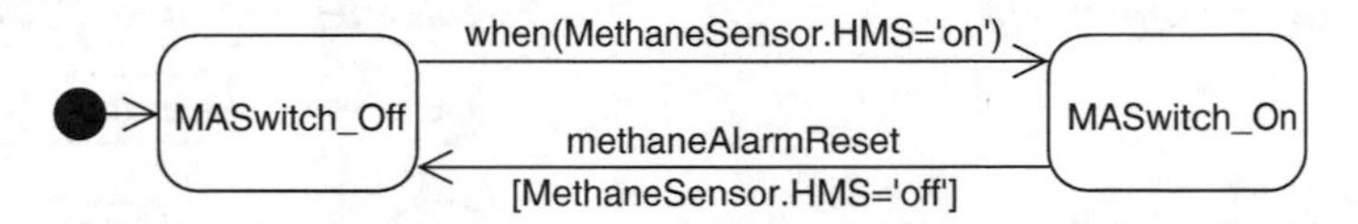

Fig. 7 State diagram of methane alarm

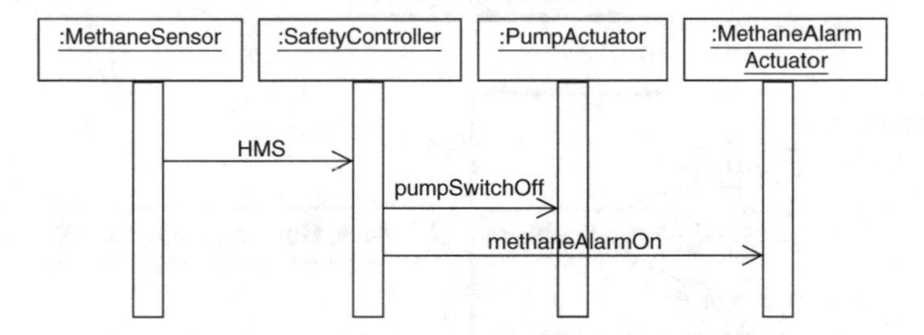

Fig. 8 A snippet of sequence diagram of MineSys

it more complete by finding more classes, attributes and relations. In our framework, this is accomplished by applying the inference rules (Figs. 4 and 5) to the current CGs Reservoir to generate new requirements in the form of CGs (Fig. 12). New requirements (shown as gray concepts and relations) are added to the CGs Reservoir.

When the CGs are transformed back to UML class diagram notations (Fig. 13), the class diagram modeler would find those new requirements (gray classes and question marks) that were not in the original diagram (Fig. 6). This generated UML classes diagram clearly provides several knowledge acquisition opportunities: attributes need to be added in the previously existing and newly generated classes, and associations need to be specified between previously existing classes and newly generated classes.

After the modeler resolves all issues in the generated class diagram, this more complete class diagram is converted to CGs again, so the CGs Reservoir becomes more complete than before. State diagrams and sequence diagrams modelers can pull their diagrams out of the central CGs Reservoir in a similar manner. That is to say, another requirements discovering and completing iteration starts.

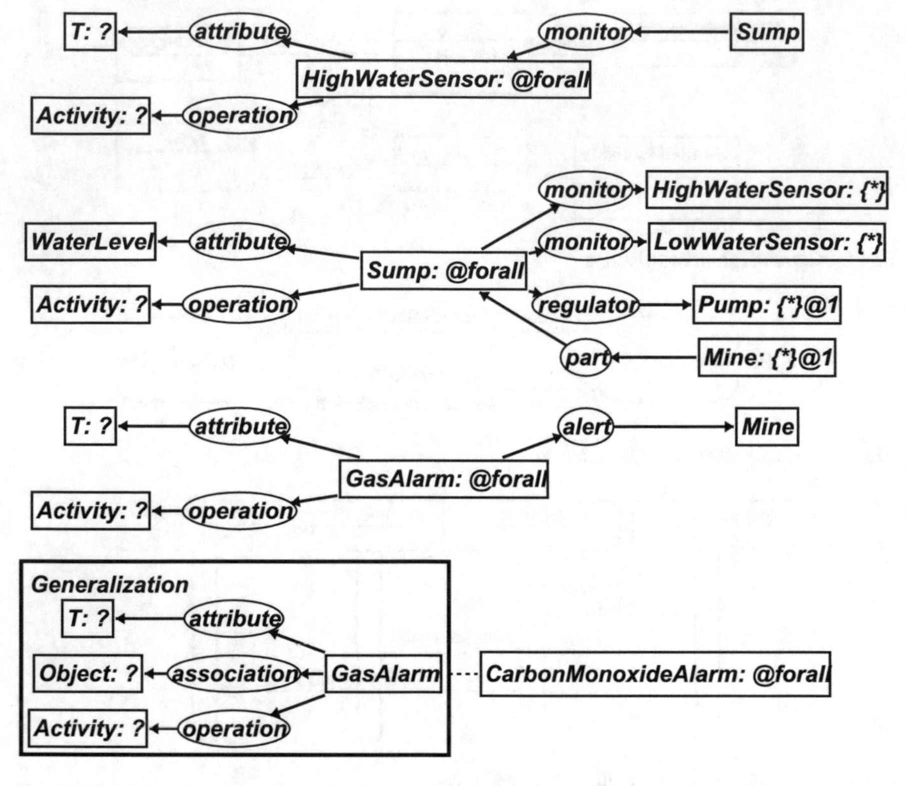

Fig. 9 The CGs of several classes in the class diagram of MineSys

4 Evaluation

In order to evaluate the effectiveness of exposing requirements acquisition opportunities using our framework, we applied it to the Mine Safety Control system and proposed one evaluation metric: The number of missing requirements that can be potentially acquired from a modeler given a generated UML diagram. This is measured by counting "semantic holes" in a generated UML diagram.

In this work, a semantic hole refers to something that needs clarification in a generated UML diagram. For example, a question mark at one end of an association or a question mark in the attribute compartment of a class or an automatically generated class name like MethaneAlarmResetMSGSender in a generated class diagram (Fig. 13). Different kinds of semantic holes that we can find for three UML diagrams are listed in Table 1. A semantic hole reveals some missing requirement that a modeler needs to provide and there is no way that a framework can generate that missing requirement automatically. An advantage of using the quantity of semantic holes as the metric is that this is objective, the number of requirements that we can get from modelers does not depend on any subjective judgment of incompleteness or

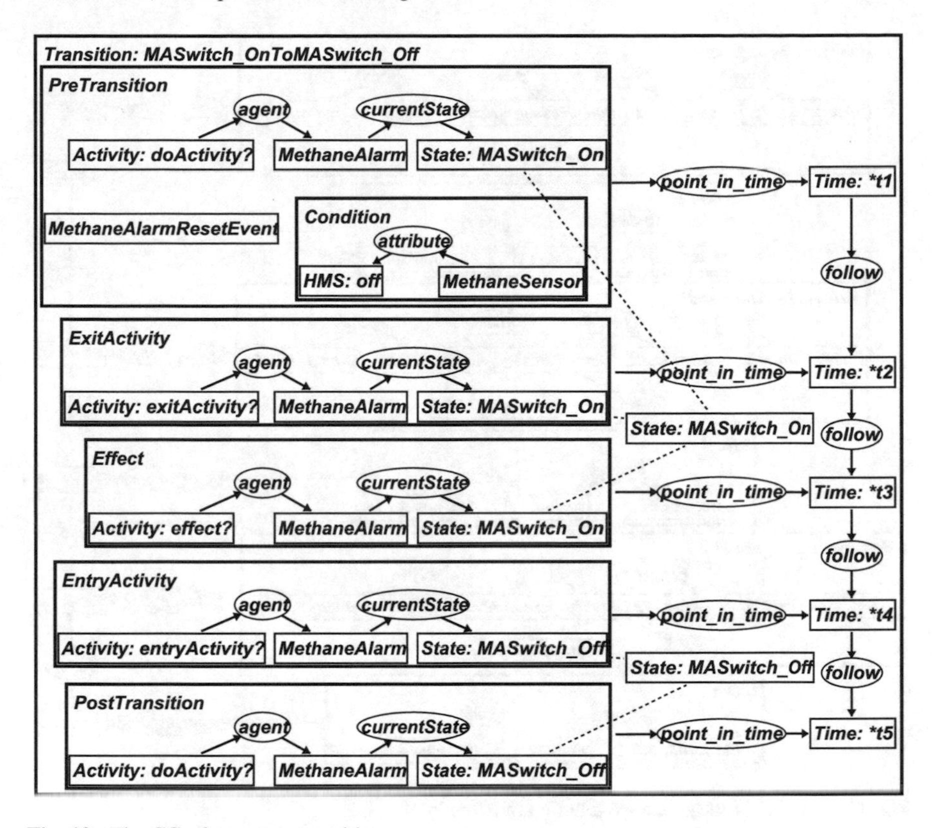

Fig. 10 The CG of one state transition

experience of requirements modelers. For example, different class diagram modelers have different ways to complete the generated class diagram in Fig. 13. So instead of asking several modelers to really fill in the holes and calculating an average, we decide to simply count the number of semantic holes that need to be filled, since these requirements (semantic holes) are missing for sure, and need to be provided by the requirements modelers.

During our evaluation process, one class diagram, four state diagrams and one sequence diagram of the MineSys were converted to CGs to populate the CGs Reservoir; based on inference rules defined in the CGs Support, new class diagram, state diagram and sequence diagrams were generated from the CGs Reservoir. Our results for the MineSys are presented in Table 1. A high number of semantic holes in a UML diagram is a sign that more requirements knowledge will be potentially acquired from a modeler. For example, in the generated class diagram, 127 semantic holes need to be resolved by the modelers.

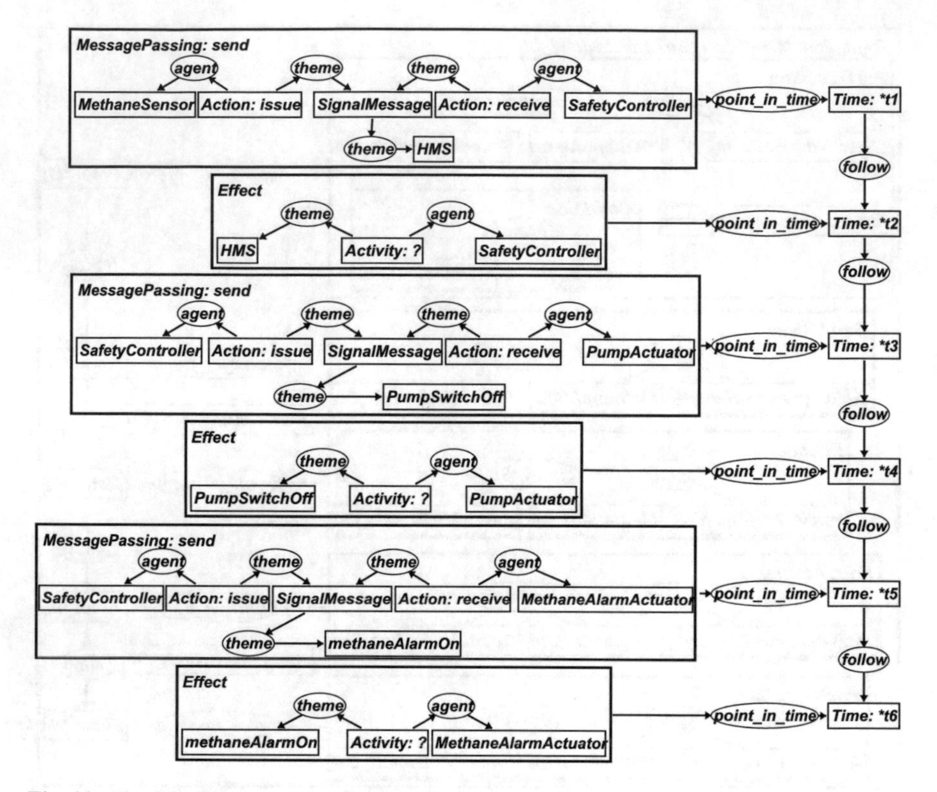

Fig. 11 The CG of the sequence diagram of MineSys

5 Discussion

In this section, we discuss several issues and the limitations of this work as well as related work and future work.

5.1 Representing Requirements in CGs

Readers may want to know more about the concepts and relations used in the canonical graphs for the three kinds of UML diagrams. Software engineering researchers have tried to identify the minimal set of fundamental elements that underlies the requirements of an object-oriented system [3, 4]. In the light of their work, the CGs Support of our framework defines a set of primitive concepts and relations underlying the three UML diagrams so that any requirement captured by the three UML diagrams can be expressed in terms of the primitives. The primitive concepts and relations in this framework are to UML as assembly language statements are to high-

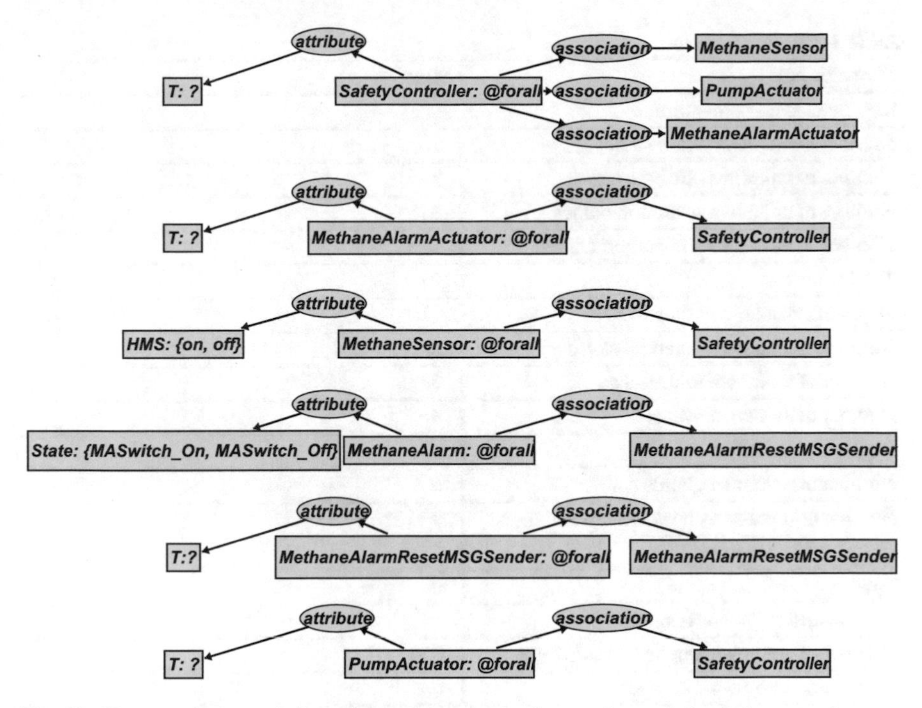

Fig. 12 New requirements inferred by applying the attribute and association inference rules

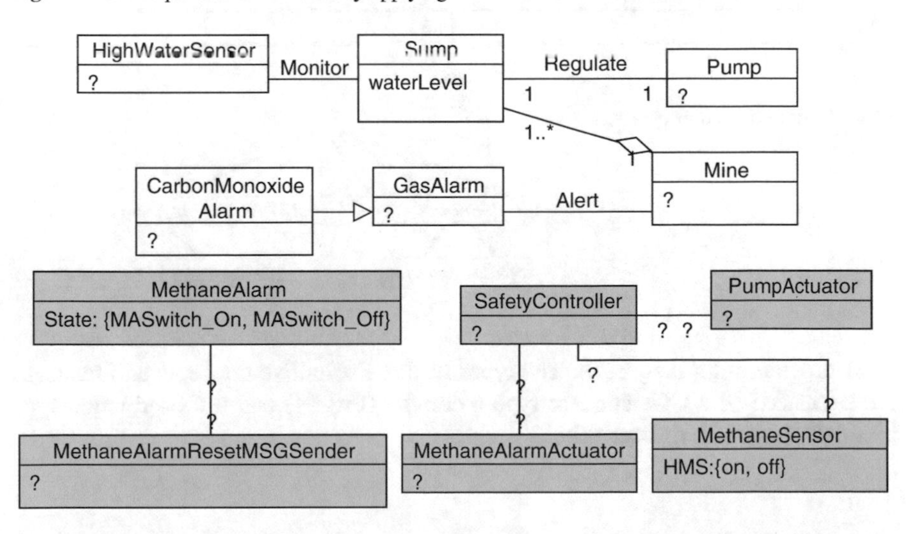

Fig. 13 Generated UML class diagram where gray represents new requirements and a '?' represents a semantic hole

Table 1 Evaluation results

Semantic holes	MineSys
In generated class diagram	
Number of unknown class names	6
Number of unknown attribute names	25
Number of unknown operation names	81
Number of unknown association names	15
Total	127
In generated state diagram	
Number of unknown/potential states	13
Number of unknown transitions	3
Number of unknown events	3
Number of unknown effects	5
Number of unknown guards	10
Number of unknown entry/exit, do activities	39
Number of state invariants	7
Total	80
In generated sequence diagram	
Number of unknown neighboring lifelines	8
Number of unknown messages	0
Number of unknown execution specs	58
Total	66

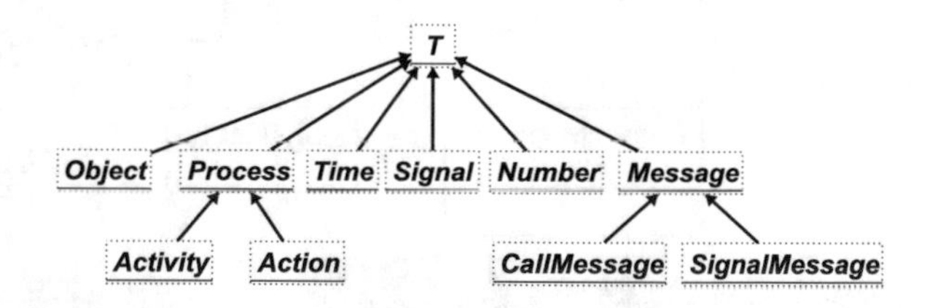

Fig. 14 Primitive concept type hierarchy

level programming languages. The types of those primitive concepts and relations are organized in a CGs concept type hierarchy (Fig. 14) and a CGs relation type hierarchy (Fig. 15), respectively.

5.2 The Size of CGs Expressing Semantics of UML Diagrams

The size of the CGs in this work is large (Look at the CG of one transition in state diagram in Fig. 10). The reason is that we are using low-level primitive concepts and relations to define each model element in a UML diagram. As a result, a semantically

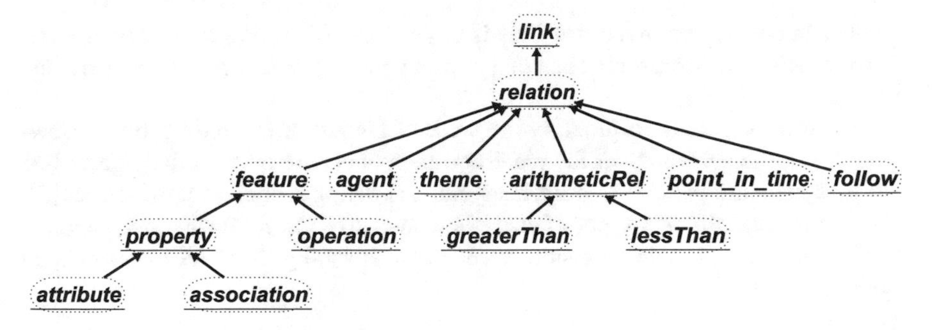

Fig. 15 Primitive relation type hierarchy

rich UML diagram takes more time and space to be expressed in primitives because each model element contains a lot of semantics. A good thing about this framework is that CGs are used as an internal representation and software requirements modelers do not have to read and understand the CGs of the three UML diagrams.

5.3 Limitations

In this work, only essential elements of the three types of UML diagrams are considered. Complex elements like association classes in class diagrams, combined fragments (loop and alt) in sequence diagrams and nested states, and concurrent states and history states in state diagrams are not supported in the framework.

Another current limitation of this work is the lack of automation support for transforming between UML and CGs. Although a CGs inference engine and a UML diagrams editor are already available [1, 13], manually converting UML diagrams to CGs and generating UML diagrams from CGs are both tedious and error-prone. Future work will focus on automating these by adopting ATL modeling transformation technique [8].

This research is still in progress. More thorough and formal evaluation with modelers involved will be carried out once the tool is available. Questionnaires and surveys with the modelers will be developed to discover their thoughts, and feelings about the tool and the transformation process. Furthermore, a comparative study is needed where we compare the effectiveness of our framework against other alternatives.

5.4 Related Work

Several previous work [2, 9, 10, 12, 14] convert requirements analysis models to a knowledge representation for the purpose of consistency checking. By contrast, our purpose in converting models to CGs is to generate UML diagrams with requirements acquisition opportunities so that more requirements can be elicited from the modelers.

Our work is greatly inspired by the work of Delugach [5] and Crye [2]. However, in [5], the semantics of the requirements models are only partially described and no systematic process of converting and generating models is provided; in [2], after converting different types of analysis models to CGs, no further discussion is provided on requirements acquisition through transforming CGs back to the analysis models.

5.5 Future Work

More UML diagrams will be included in our framework and we plan to develop a web-based system adopting the framework which facilitates requirements acquisition process among a team of requirements modelers preparing a software specification in different views. Through the web, the team members can get quick feedback from other models. Also, a real industrial-strength example will be used in the framework for further evaluation.

6 Conclusion

In this work, a requirements acquisition framework using UML and conceptual graphs is developed. By transforming UML diagrams to and from CGs, requirements acquisition opportunities are exposed. This framework is useful for the requirements acquisition process among a team of requirements modelers preparing a software specification from different viewpoints. Our framework has been successfully used in a case study; the results and evaluations have shown the effectiveness of our framework in facilitating modelers in acquiring requirements.

References

1. CoGui-Lirmm: A Conceptual Graph Editor. http://www.lirmm.fr/cogui/. Accessed 18 Mar 2016.
2. Cyre, W. R. (1997). Capture, integration, and analysis of digital system requirements with conceptual graphs. *IEEE Transactions on Knowledge and Data Engineering*, *9*(1), 8–23.
3. Dardenne, A., Van Lamsweerde, A., & Fickas, S. (1993). Goal-directed requirements acquisition. *Science of Computer Programming*, *20*(1), 3–50.
4. Davis, A. M., Jordan, K., & Nakajima, T. (1997). Elements underlying the specification of requirements. *Annals of Software Engineering*, *3*(1), 63–100.

5. Delugach, H. S. (1996). An approach to conceptual feedback in multiple viewed software requirements modeling. In *Joint Proceedings of the Second International Software Architecture Workshop (ISAW-2) and International Workshop on Multiple Perspectives in Software Development (Viewpoints' 96) on SIGSOFT'96 Workshops* (pp. 242–246).
6. Firesmith, D. (2005). *Journal of Object Technology*, *4*(1), 27–44.
7. Jaramillo, C. M. Z., Gelbukh, A., & Isaza, F. A. (2006). Pre-conceptual schema: A conceptual-graph-like knowledge representation for requirements elicitation. In *MICAI 2006: Advances in Artificial Intelligence* (pp. 27–37).
8. Jouault, F., Allilaire, F., & Bzivin, J., Kurtev, I. (2008). ATL: A model transformation tool. *Science of Computer Programming*, *72*(1), 31–9.
9. Lucas, F. J, Molina, F., & Toval, A. A systematic review of UML model consistency management. *Information and Software Technology*, *51*(12), 1631–45.
10. Shan, L., & Zhu, H. (2008). A formal descriptive semantics of UML. *Formal methods and software engineering* (pp. 375–396). Heidelberg: Springer.
11. Sowa, J. F. (1983). *Conceptual structures: Information processing in mind and machine*. Reading, MA: Addison-Wesley Publication.
12. Sunetnanta, T., & Finkelstein, A. (2001). Automated consistency checking for multiperspective software specifications. In *Workshop on Advanced Separation of Concerns*. Toronto.
13. UMLet—Free UML Tool for Fast UML Diagrams. http://www.umlet.com/. Accessed 18 Mar 2016.
14. Van Der Straeten, R., Mens, T., Simmonds, J., & Jonckers, V. (2003). Using description logic to maintain consistency between UML models. In *UML 2003-The Unified Modeling Language*.
15. Van Lamsweerde, A. (2009). *Requirements engineering: From system goals to UML models to software specifications*. England: Wiley.
16. Wei B (2015) A comparison of two frameworks for multiple-viewed software requirements acquisition. Dissertation, University of Alabama in Huntsville.

Identification Method of Fault Level Based on Deep Learning for Open Source Software

Yoshinobu Tamura, Satoshi Ashida, Mitsuho Matsumoto and Shigeru Yamada

Abstract Recently, many open source software are used for quick delivery, cost reduction, standardization. The bug tracking systems are managed by many open source projects. Then, many data sets are recorded on the bug tracking systems by many users and project members. The quality of open source software will be improved significantly if the software managers can make an effective use of these data sets on the bug tracking systems. In this paper, we propose a method of open source software reliability assessment based on the deep learning. Also, we show several numerical examples of open source software reliability assessment in the actual software projects. Moreover, we compare the methods to estimate the level of software faults based on the deep learning by using the fault data sets of actual software projects.

1 Introduction

Various open source software (OSS) have been developed under many open source projects. However, the poor handling of quality problem prohibits the progress of OSS, because the development cycle of OSS has no specified testing-phase. In particular, the bug tracking systems are used in many open source projects. Many fault data sets are recorded on these bug tracking system. The quality of open source soft-

Y. Tamura (✉) · S. Ashida
Yamaguchi University, Tokiwadai 2-16-1, Ube-shi, Yamaguchi 755-8611, Japan
e-mail: tamura@yamaguchi-u.ac.jp

S. Ashida
e-mail: v002vk@yamaguchi-u.ac.jp

M. Matsumoto · S. Yamada
Tottori University, Minami 4-101, Koyama, Tottori-shi 680-8552, Japan
e-mail: M15T7019Y@edu.tottori-u.ac.jp

S. Yamada
e-mail: yamada@sse.tottori-u.ac.jp

R. Lee (ed.), *Software Engineering Research, Management and Applications*, Studies in Computational Intelligence 654,
DOI 10.1007/978-3-319-33903-0_5

ware will be improved significantly if the software managers can make an effective use of these data sets on the bug tracking systems.

In the past, many software reliability models [1–3] have been applied to assess the reliability for quality management and testing-progress control of software development. However, it is difficult for the software managers to select the optimal software reliability model for the actual software development project. As an example, the software managers can assess the software reliability for the past data sets by using the model evaluation criteria. On the other hand, the estimation results based on the past fault data cannot be guaranteed for the future data sets of actual software projects. Therefore, it is difficult for the software managers to assess the reliability of OSS by using the SRGM's. Moreover, the software managers will be need to convert the fault data on bug tracking system from the raw data to the fault count data. It is efficient compared with the conventional method based on SRGM's if the software managers can use all raw data of bug tracking system.

In this paper, we focus on the identification method of software fault level. Then, we propose the method of OSS reliability assessment based on deep learning. Also, several numerical examples of software reliability assessment by using the fault data in the actual OSS projects are shown. Moreover, we compare the methods to estimate the cumulative numbers of detected faults based on the deep learning with that based on neural network.

2 Identification Method of Software Fault Level Based on Neural Network

The structure of the neural networks in this paper is shown in Fig. 1. Let $w^1_{ij}(i = 1, 2, \dots, I; j = 1, 2, \dots, J)$ be the connection weights from i-th unit on the sensory layer to j-th unit on the association layer, $w^2_{jk}(j = 1, 2, \dots, J; k = 1, 2, \dots, K)$ denote the connection weights from j-th unit on the association layer to k-th unit on the response layer. Moreover, $x_i(i = 1, 2, \dots, I)$ represent the normalized input values of i-th unit on the sensory layer, and $y_k(k = 1, 2, \dots, K)$ are the output values. We apply the actual number of detected faults per unit time $N_i(i = 1, 2, \dots, I)$ to the input values $x_i(i = 1, 2, \dots, I)$.

Considering the amount of characteristics for the software fault data on bug tracking systems, we apply the following amount of information as parameters to the input data $x_i(i = 1, 2, \dots, I)$.

- Date recorded on bug tracking system
- Name of software product
- Name of software component
- Number of software version
- Nickname of fault reporter
- Nickname of fault assignee
- Status of software fault
- Name of operating system

Fig. 1 The structure of our neural network based on back-propagation

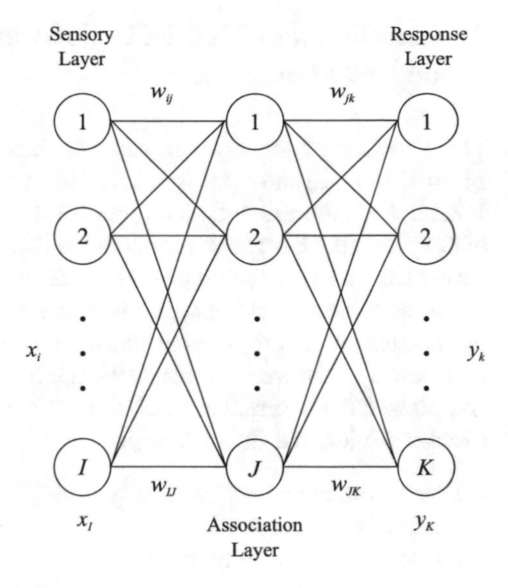

The input-output rules of each unit on each layer are given by

$$h_j = f\left(\sum_{i=1}^{I} w_{ij}^{1} x_i\right), \tag{1}$$

$$y_k = f\left(\sum_{j=1}^{J} w_{jk}^{2} h_j\right), \tag{2}$$

where a logistic activation function $f(\cdot)$ which is widely-known as a sigmoid function given by the following equation:

$$f(x) = \frac{1}{1 + e^{-\theta x}}, \tag{3}$$

where θ is the gain of sigmoid function. We apply the multi-layered neural networks by back-propagation in order to learn the interaction among software components [4]. We define the error function given by the following equation:

$$E = \frac{1}{2}\sum_{k=1}^{K}(y_k - d_k)^2, \tag{4}$$

where $d_k (k = 1, 2, \ldots, K)$ are the target input values for the output values. We apply 8 kinds of fault level to the amount of compressed characteristics, i.e., Trivial, Enhancement, Minor, Normal, Regression, Blocker, Major, and Critical, respectively. Then, the number of units K in response layer is 8 because of 8 fault levels.

3 Identification Method of Software Fault Level Based on Deep Learning

The structure of the deep learning in this paper is shown in Fig. 2. In Fig. 2, $z_l(l = 1, 2, \ldots, L)$ and $z_m(m = 1, 2, \ldots, M)$ means the pre-training units. Also, $o_n(n = 1, 2, \ldots, N)$ is the amount of compressed characteristics. Several algorithms in terms of deep learning have been proposed [5–10]. In this paper, we apply the deep neural network to learn the fault data on bug tracking systems of open source projects.

As with the neural network, we apply the following amount of information to the parameters of pre-training units. Then, the objective variable is given as the fault levels as shown in Table 1. We apply 8 kinds of fault level to the amount of compressed characteristics, i.e., Trivial, Enhancement, Minor, Normal, Regression, Blocker, Major, and Critical, respectively.

- Date recorded on bug tracking system
- Name of software product
- Name of software component

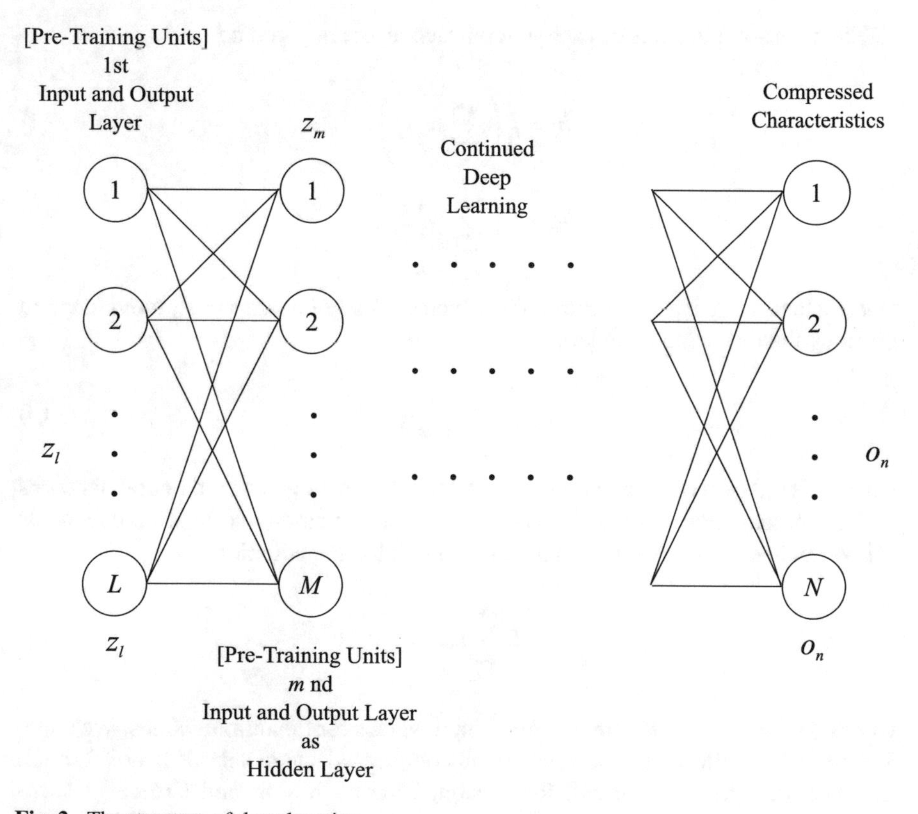

Fig. 2 The structure of deep learning

Table 1 The fault levels in learning data

Index number	Fault level
1	Trivial
2	Enhancement
3	Minor
4	Normal
5	Regression
6	Blocker
7	Major
8	Critical

- Number of software version
- Nickname of fault reporter
- Nickname of fault assignee
- Status of software fault
- Name of operating system

8 kinds of explanatory variables are set to the amount of pre-training units. Then, each data of explanatory variable is converted from the character data to the numerical value such as the rate of occurrence.

4 Numerical Examples

The OSS is closely watched from the point of view of the cost reduction and the quick delivery. There are several open source projects in the area of server software. In particular, we focus on Apache HTTP server [11] in order to evaluate the performance of our methods. In this paper, we show numerical examples by using the data sets for Apache HTTP server as OSS. The data used in this paper are collected in the bug tracking system on the website of Apache HTTP server open source project. We show the raw data obtained from the bug tracking system in Fig. 3. Also, Fig. 4 is the input data for deep learning. Figure 4 can be obtained by converting from the character data to the numerical value such as the rate of occurrence.

	A	B	C	D	E	F	G	H	I
1	Opened	Product	Component	Version	Reporter	Assignee	Status	OS	Severity
2	2000/8/26 5:48	Tomcat 3	Unknown	Unknown	dwd	dev	RESOLVED	All	normal
3	2000/8/28 13:56	Tomcat 3	Connectors	3.1 Final	anonymous-bug	dev	RESOLVED	All	normal
4	2000/8/28 15:10	Tomcat 3	Jasper	3.1 Final	matthias.ernst	dev	RESOLVED	All	normal
5	2000/8/28 18:54	Tomcat 3	Unknown	Unknown	geoff	dev	RESOLVED	All	normal
6	2000/8/29 13:57	Tomcat 3	Connectors	3.1 Final	anonymous-bug	dev	RESOLVED	All	normal
7	2000/8/29 15:04	Tomcat 3	Jasper	3.1 Final	anonymous-bug	dev	RESOLVED	All	normal
8	2000/8/30 3:59	Tomcat 3	Auth	3.1 Final	alex.den.heijer	dev	RESOLVED	All	normal
9	2000/8/30 17:36	Tomcat 3	Connectors	3.2.1 Final	bren	dev	RESOLVED	All	normal
10	2000/8/31 6:45	Tomcat 3	Unknown	3.1 Final	anonymous-bug	dev	RESOLVED	All	normal
11	2000/8/31 12:54	Tomcat 3	Servlet	3.2.x Nightly	anonymous-bug	dev	RESOLVED	All	normal
12	2000/9/1 19:28	Tomcat 3	Connectors	3.1 Final	jean-paul_abgrall	dev	RESOLVED	All	normal
13	2000/9/2 17:31	Tomcat 3	Jasper	3.1 Final	todd	dev	RESOLVED	All	normal
14	2000/9/4 8:38	Tomcat 3	Servlet	3.2.1 Final	Laurent.Salle	dev	RESOLVED	All	normal
15	2000/9/5 5:59	Tomcat 3	Servlet	3.1 Final	lamberto	dev	RESOLVED	All	normal
16	2000/9/7 3:38	Tomcat 3	Jasper	3.2.1 Final	anonymous-bug	dev	RESOLVED	All	normal
17	2000/9/7 8:36	Tomcat 3	Jasper	3.2.1 Final	iblesa	dev	RESOLVED	All	normal
18	2000/9/7 20:20	Ant	Core tasks	1.2	anand	notifications	CLOSED	All	normal
19	2000/9/7 20:22	Tomcat 3	Connectors	3.2.1 Final	anand	dev	RESOLVED	All	normal
20	2000/9/8 3:03	Tomcat 3	Auth	3.1.1 Final	maxom	dev	RESOLVED	All	normal

Fig. 3 A part of the raw data obtained from the bug tracking system

	A	B	C	D	E	F	G	H	I
1	opened	product	component	version	reporter	assignee	status	os	severity
2	2.338831019	0.051	0.024	0.0042	0.0114	0.4981	0.7856	0.380254841	normal
3	0.051087963	0.051	0.0346	0.0042	0.0001	0.4981	0.7856	0.380254841	normal
4	0.155914352	0.051	0.0269	0.0128	0.0003	0.4981	0.7856	0.380254841	normal
5	0.793391204	0.051	0.024	0.0042	0.0114	0.4981	0.7856	0.380254841	normal
6	0.046527778	0.051	0.0346	0.0042	0.0114	0.4981	0.7856	0.380254841	normal
7	0.538113426	0.051	0.0017	0.0042	0.0001	0.4981	0.7856	0.380254841	normal
8	0.567662037	0.051	0.024	0.0277	0.0001	0.4981	0.7856	0.380254841	normal
9	0.547986111	0.051	0.0269	0.0042	0.0114	0.4981	0.7856	0.380254841	normal
10	0.255949074	0.051	0.0135	0.005	0.0114	0.4981	0.7856	0.380254841	normal
11	1.27369213	0.051	0.024	0.0042	0.0001	0.4981	0.7856	0.380254841	normal
12	0.918715278	0.051	0.0346	0.0042	0.0002	0.4981	0.7856	0.380254841	normal
13	1.630289352	0.051	0.0135	0.0277	0.0001	0.4981	0.7856	0.380254841	normal
14	0.889618056	0.051	0.0135	0.0042	0.0001	0.4981	0.7856	0.380254841	normal
15	1.901689815	0.051	0.0346	0.0277	0.0114	0.4981	0.7856	0.380254841	normal
16	0.207071759	0.051	0.0346	0.0277	0.0001	0.4981	0.7856	0.380254841	normal
17	0.488865741	0.0899	0.0367	0.0215	0.0002	0.0923	0.0809	0.380254841	normal
18	0.001689815	0.051	0.024	0.0277	0.0002	0.4981	0.7856	0.380254841	normal
19	0.278101852	0.051	0.0017	0.0024	0.0001	0.4981	0.7856	0.380254841	normal
20	0.156076389	0.051	0.0135	0.0277	0.0027	0.4981	0.7856	0.380254841	normal

Fig. 4 A part of the input data for deep learning

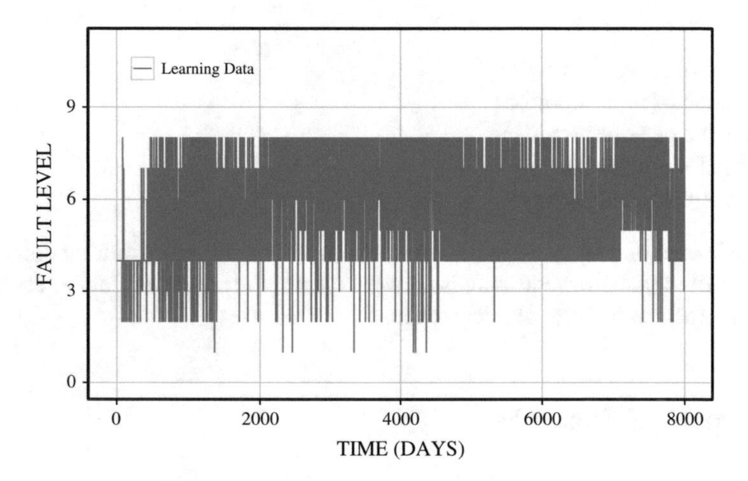

Fig. 5 The learning data recorded on bug tracking system

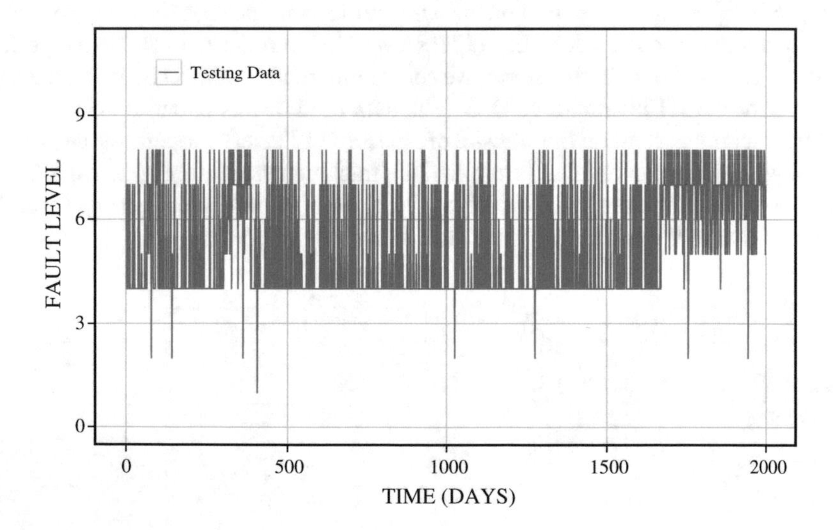

Fig. 6 The testing data recorded on bug tracking system

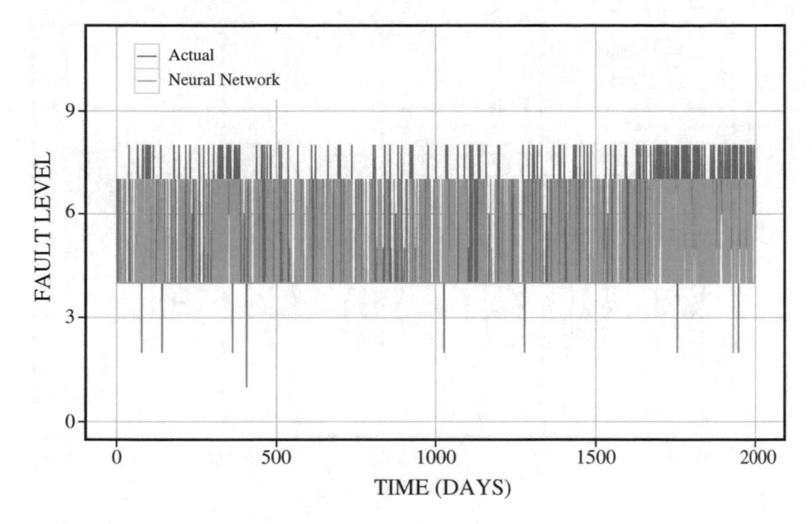

Fig. 7 The estimation results of fault levels based on neural network

We obtain 10,000 fault data set from the data recorded on bug tracking system of Apache HTTP server. Then, 80 % of the recorded data is used as the learning data. Figure 5 shows the learning data. The fault levels of Fig. 5 are shown in Table 1. We show the estimation results by using the testing data sets in Fig. 6.

4.1 Estimation Results

We apply 8 fault levels as the fault levels on bug tracking systems to the objective variable. The estimation results for 2,000 testing data set based on neural network by using 8,000 learning data set is shown in Fig. 7. Similarly, the estimation results for 2,000 testing data set based on deep learning by using 8,000 learning data set is shown in Fig. 8. Moreover, we show the estimation results based on neural network and deep learning in Tables 2 and 3.

From Figs. 7 and 8, we can confirm that the estimate based on deep learning fits better than one based on neural network for the future in fact.

4.2 Comparison Results

The estimated results of recognition rate based on the neural network and deep learning are shown in Table 4. From Table 4, we found that the estimated recognition rates based on the deep learning perform better than that of the neural network.

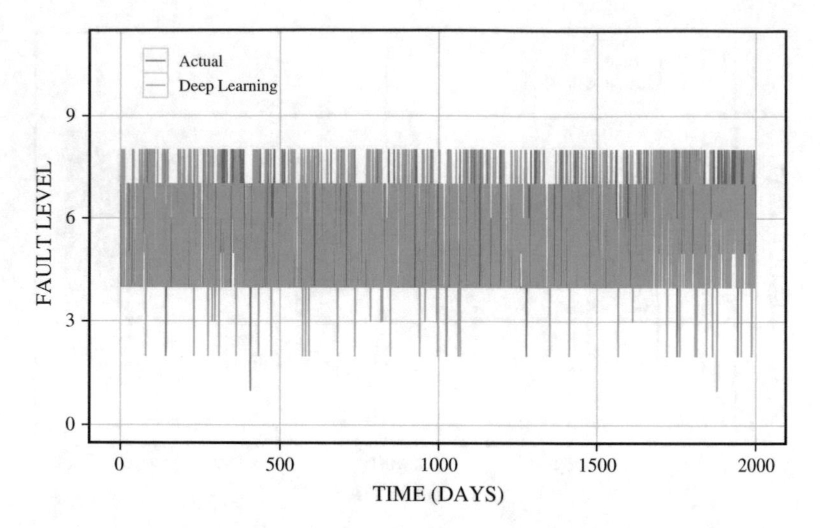

Fig. 8 The estimation results of fault levels based on deep learning

Table 2 The estimation results based on neural network

Estimate	Blocker	Critical	Enhancement	Major	Minor	Normal	Regression	Trivial
Blocker	0	0	0	0	0	0	0	0
Critical	0	0	0	0	0	0	0	0
Enhancement	0	0	0	1	0	0	0	0
Major	57	87	2	146	0	190	35	0
Minor	0	0	0	0	0	0	0	0
Normal	55	121	5	246	0	988	65	1
Regression	0	0	0	0	0	0	0	0
Trivial	0	0	0	0	0	0	0	0

Table 3 The estimation results based on deep learning

Estimate	Blocker	Critical	Enhancement	Major	Minor	Normal	Regression	Trivial
Blocker	5	14	0	24	0	47	8	0
Critical	22	39	0	57	0	125	16	0
Enhancement	2	5	2	5	0	15	1	0
Major	47	71	1	128	0	315	33	1
Minor	0	0	0	0	0	8	0	0
Normal	33	73	4	166	0	645	41	0
Regression	3	6	0	13	0	23	0	0
Trivial	0	0	0	0	0	0	1	0

Table 4 The comparison results for methods based on neural network and deep learning

Estimation method	Recognition rate (%)
Neural network	4.6023
Deep learning	40.970

Table 5 The fault levels in learning data in case of two categories

Index number	Fault level
1	Other (Trivial)
1	Other (Enhancement)
1	Other (Minor)
1	Other (Normal)
1	Other (Regression)
1	Other (Blocker)
2	Major (Major)
2	Major (Critical)

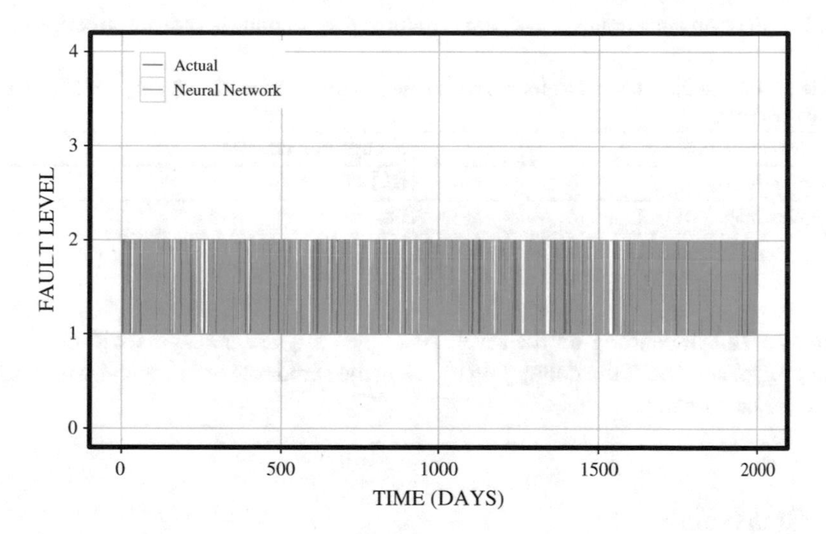

Fig. 9 The estimation results of fault levels based on neural network in case of two categories

Moreover, we assess the effectiveness in case of two kinds of fault level considering the practicality of the proposed method. Then, we consider the case of Table 5. Figures 9 and 10 show the estimation results for 2,000 testing data set based on neural network and deep learning by using 8,000 learning data set, respectively. Table 6 shows the comparison results for methods based on neural network and deep learning in case of two categories. In particular, the estimation results based on the deep learning give a high-recognition rates. It will be possible for the software managers to predict the fault level of importance from the fault data recorded on bug tracking system. The proposed method will be useful to make a quick modification of

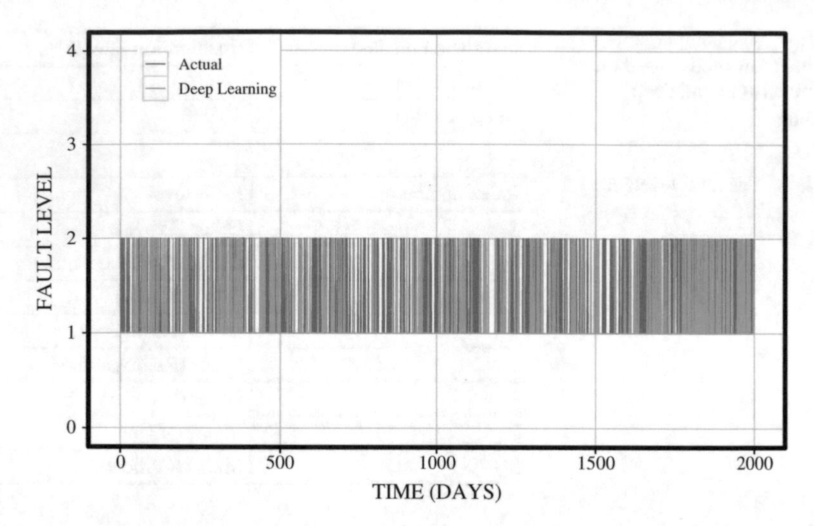

Fig. 10 The estimation results of fault levels based on deep learning in case of two categories

Table 6 The comparison results for methods based on neural network and deep learning in case of two categories

Estimation method	Recognition rate (%)
Neural network	65.183
Deep learning	68.584

software fault depending on the level. Also, the software managers will be able to take prompt action for the debugging by using the standards of judgment with regard to major fault or not.

5 Conclusion

At present, the bug tracking systems are used in many open source projects. Then, many fault data sets are recorded on these bug tracking system. The quality of open source software will be improved significantly if the software managers can make an effective use of these data sets on the bug tracking systems. In case of using the bug tracking systems, the software managers will be able to take prompt action for the debugging process, if the software managers can judge with regard to major fault or not.

This paper have focused on the identification method of software fault level. We have proposed the method of reliability assessment based on deep learning. In particular, it is difficult to judge the major fault by only using the data on bug tracking system, because the contents of data recorded on bug tracking system cannot be guaranteed for the results occurred from the actual OSS, i.e., many general users as well as the major project member can report on the bug tracking system. Also, several numerical examples of OSS reliability assessment by using the fault data in the actual OSS project have been shown in this paper. Moreover, we have compared the estimation method based on the deep learning with that based on neural network. Thereby, we have found that our method can assess OSS reliability in the future with high accuracy based on the data on bug tracking system.

In the future study, it will be necessary to analyze by using many training data sets in actual software development projects. Thereby, the proposed method based on the deep learning will be useful for the software managers to assess the OSS reliability.

Acknowledgments This work was supported in part by the Telecommunications Advancement Foundation in Japan, and the JSPS KAKENHI Grant No. 15K00102 and No. 25350445 in Japan.

References

1. Lyu, M. R. (Ed.). (1996). *Handbook of software reliability engineering*. Los Alamitos, CA: IEEE Computer Society Press.
2. Yamada, S. (2014). *Software reliability modeling: Fundamentals and applications*. Heidelberg: Springer.
3. Kapur, P. K., Pham, H., Gupta, A., & Jha, P. C. (2011). *Software reliability assessment with or applications*. London: Springer.
4. Karnin, E. D. (1990). A simple procedure for pruning back-propagation trained neural networks. *IEEE Transactions on Neural Networks*, *1*, 239–242.
5. Kingma, D. P., Rezende, D. J., Mohamed, S., & Welling, M. (2014). Semi-supervised learning with deep generative models. *Proceedings of Neural Information Processing Systems*.
6. Blum, A., Lafferty, J., Rwebangira, M. R., & Reddy, R. (2004). Semi-supervised learning using randomized mincuts. *Proceedings of the International Conference on Machine Learning*.
7. George, E. D., Dong, Y., Li, D., & Alex, A. (2012). Context-dependent pre-trained deep neural networks for large-vocabulary speech recognition. *IEEE Transactions on Audio, Speech, and Language Processing*, *20*(1), 30–42.
8. Vincent, P., Larochelle, H., Lajoie, I., Bengio, Y., & Manzagol, P. A. (2010). Stacked denoising autoencoders: Learning useful representations in a deep network with a local denoising criterion. *Journal of Machine Learning Research*, *11*(2), 3371–3408.
9. Martinez, H. P., Bengio, Y., & Yannakakis, G. N. (2013). Learning deep physiological models of affect. *IEEE Computational Intelligence Magazine*, *8*(2), 20–33.
10. Hutchinson, B., Deng, L., & Yu, D. (2013). Tensor deep stacking networks. *IEEE Transactions on Pattern Analysis and Machine Intelligence*, *35*(8), 1944–1957.
11. The Apache Software Foundation, The Apache HTTP Server Project. http://httpd.apache.org/.

Monitoring Target Through Satellite Images by Using Deep Convolutional Networks

Xudong Sui, Jinfang Zhang, Xiaohui Hu and Lei Zhang

Abstract Monitoring target through satellite images is widely used in intelligence analysis and for anomaly detection. Meanwhile, it is also challenging due to the shooting conditions and the huge amounts of data. We propose a method for target monitoring based on deep convolutional neural networks (DCNN). The method is implemented by three procedures: (i) Label the target and generate the dataset, (ii) train a classifier, and (iii) monitor the target. First, the target area is labelled manually to form a dataset. In the second stage a classifier based on DCNN using Keras library is well-trained. In the last stage the target is monitored in the test satellite images. The method was tested on two different application scenarios. The results show that the mothed is effective.

Keywords Target monitoring · DCNN · Intelligence analysis · Abnormal weather detection

1 Introduction

With the rapid technological development of various satellites, a huge volume of high-resolution images can now be obtained easily from Google Earth Pro, which also provides historical images in many areas. Such development makes it possible

X. Sui (✉) · J. Zhang · X. Hu
Institute of Software Chinese Academy of Sciences, Beijing, China
e-mail: xudong2014@iscas.ac.cn

J. Zhang
e-mail: jinfang@iscas.ac.cn

X. Hu
e-mail: hxh@iscas.ac.cn

L. Zhang
School of Computer Science, Beijing Information Science & Technology University, Beijing 100101, China
e-mail: 398173551@qq.com

R. Lee (ed.), *Software Engineering Research, Management and Applications*, Studies in Computational Intelligence 654,
DOI 10.1007/978-3-319-33903-0_6

for target monitoring in satellite image series, which can be used in intelligence analysis, anomaly detection, and various other purposes. Xi Li et al. [1] used the night-time light images to evaluate the Syrian crisis and it has been paid attention in the international society. Yet affected by the shooting conditions such as illumination, clouds, fog, noise and other factors, target monitoring is a challenging job. Meanwhile, the huge amounts of satellite remote sensing images also cost researchers much more time to do the job. Relying on manual work is extremely time-consuming and costly. Therefore, there is a need to automate the process of monitoring target.

Monitoring problem can be treated as a detection problem in image series. In general, a well-trained classifier is needed to detect. The CNN [2], which was first proposed by LeCun et al. In 2012, Krizhevsky et al. [3] demonstrated that convolutional neural networks (CNNs) obtained substantially higher image classification accuracy on the ImageNet Large Visual Recognition Challenge (ILSVRC). From then on, a classifier based on DCNN [3–5] is the best choice in image processing area. The features like colors, shape, appearance of different targets are different in different application scenarios. Design feature manually is a giant project. However, we don't need to design features by hand using DCNN. So we choose the DCNN as the classifier of the monitor. Features extracted by using DCNN and then used by a classifier can gain a high accuracy. Salberg [6] used it to detect seals in remote sensing images and got an accuracy of 98.2 %. Wu et al. [7] used it to detect planes and gained a fast and accurate result. However, their methods handle only one image from a specific moment, which is not enough. Chen et al. [8] developed a deep tracking system, which inspired us. The GPU has developed a lot and the GPGPU technology is easier to use in these few years [9]. Chollet [10] is a deep learning library using GPU to accelerate, which inspired us to apply the monitoring job with GPU. We use a GTX Titan X GPU card to accelerate the job. It can save much time on monitoring job.

In this paper, we issued a method based on DCNN using Keras library. After labeling the target using Photoshop by hand, the system can automatically do the monitoring. We will test the mothed for a military target scenario and an abnormal weather detection scenario.

The rest of the paper is organized as follows. In Sect. 2, we describe our method in details. Details of experiments and results are presented in Sect. 3. Section 4 concludes the paper.

2 Method

To monitor target in satellite image series accurately and efficiently, the classifier is the key. So we consult some classic DCNN architecture [2–5]. The proposed target monitoring method consists of three stages: Label and generate the dataset, train a classifier, monitor the target. Figure 1 shows the pipeline of our target monitoring system.

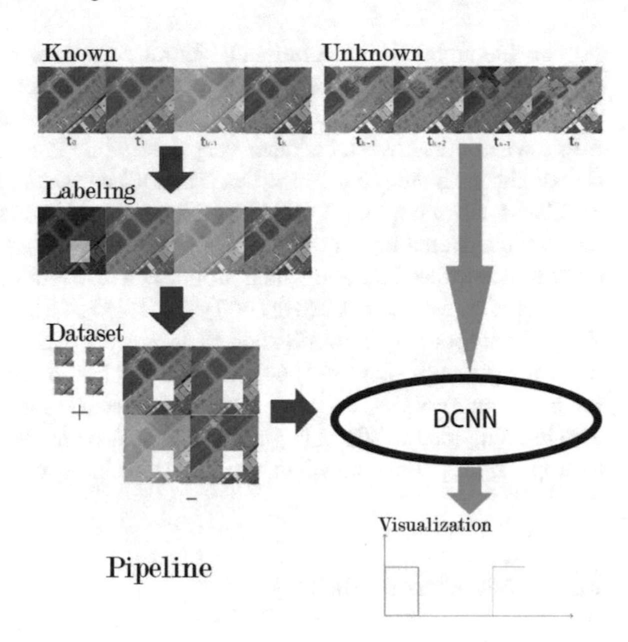

Fig. 1 Target monitoring pipeline

2.1 *Label and Generate the Dataset*

We have n satellite image series with no label, k of n images are after registration and will be used for training. The left images will be used for test.

Around the area of interest, we use two different colors to label the target area and the background respectively. We label the target area by white color and label the background by black color. Save the label layer as label image L. Consider a 3×3 grids, the center is the target and the others are background. So the negative ones (background area) is 8 times of the positive ones (target area).

Label image L is matched well to any one of the satellite image S from time t_0 to time t_k, so the class of one sample in any coordinates (x, y) from L is as same as the one in the coordinate (x, y) from any one of the satellite image S. Then we get the training dataset.

To make sure the classifier based on DCNN works well, we need to prepare enough positive and negative samples. If we did not get enough samples in the last step, we can random rotate the samples with a small angle (less than 2°) to generate enough samples.

2.2 *Architecture of DCNN*

A typical DCNN is a multilayer architecture which stacks convolution and pooling layers in alternation, with a fully connected layer to produce the final result of the task. In classification tasks, each unit of the final layer indicates the class probability.

The hierarchical Convolutional Neural Networks model considers high level image representation. However, the more layers, the more parameters in the networks needed to learn and more time needed to train. And it's more easily to become overfitting when the network becomes very deep while the training data is not sufficient. So we choose a simple architecture. The DCNN architecture we use is: C1-S2-C3-S4-C5-S6-F7, where C, S and F represent convolutional layer, max pooling layer and full connected layer respectively. We feed the output of the last fully-connected layer to a softmax function which produces a distribution over the class labels. The configuration we use is C1(8@7 × 7), C3(16@5 × 5), C5(24@3 × 3) and F7(512). We put a dropout layer after S6 and F7 to prevent overfitting. The input of DCNN is RGB images with size (64 × 64). We use Stochastic gradient descent, with support for momentum = 0.9, decay = 1e-6, and Nesterov momentum to train the model. The learning rate is 0.01 at beginning. The batch size is 128. We use early-stopping to stop training when the validation loss stops improving with patience 2.

2.3 *Monitoring Job*

Because image from t_{k+1} to t_n is not registered and the target is randomly drift less than 20 m due to the GPS error and other reasons, we have to resample from the area around. We use the coordinates information we got in the first step to reduce the search space.

Then we give the samples to the classifier we trained last stage. We set a threshold of 0.989 to get the right activation, that is, the sample is more like a positive one. If the class is positive, we mark it with red color and show it in the picture.

We apply the classifier to every test image frame and will get an events flow. The visualization result will tell the changes (exist or not) of the target in the image series.

3 Experiment and Results

3.1 *Dataset*

We will test the method for a scenario of military target (used for intelligence analysis) and a scenario of abnormal weather detection case. These two scenarios actually provide two different datasets (airplane and buildings). We have 6 parking apron images and 20 building images from different time downloaded from Google Earth Pro with resolution 1 and 0.5 m/pixel respectively. 6 parking apron images are not enough but Google cannot provide more full images of the area shot from the same time due to unknown reasons.

3.2 Military Intelligence Analysis Scenarios

We pick 3 images containing an airplane in target area. First, label one airplane with white color and label the background with black color at one time. Figure 2 shows the fusion effects of the satellite image series from 3 moments and part of the label layer.

The sample size is 64×64. Figure 3 shows part of the positive samples and negative samples. Generally, we have to rotate the positive samples with small angles (less than 2°) to get more (data augmentation [3]). Then we get the training dataset (1998 positive samples and 7998 negative ones, 90 % is from data augmentation). 20 % of the data will be used as validation data.

The classifier can achieve training accuracy at almost 98 % with validation accuracy at 98 % on the dataset. We have tested the code provided by the Keras' author

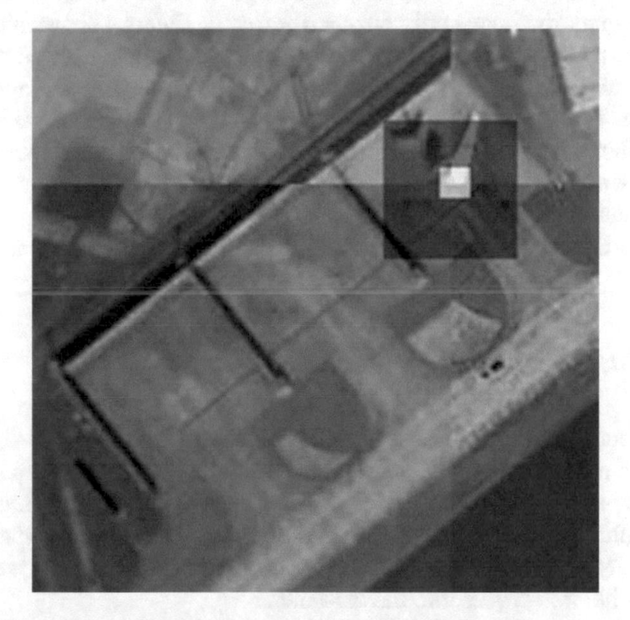

Fig. 2 The fusion of satellite image series and part of the label image

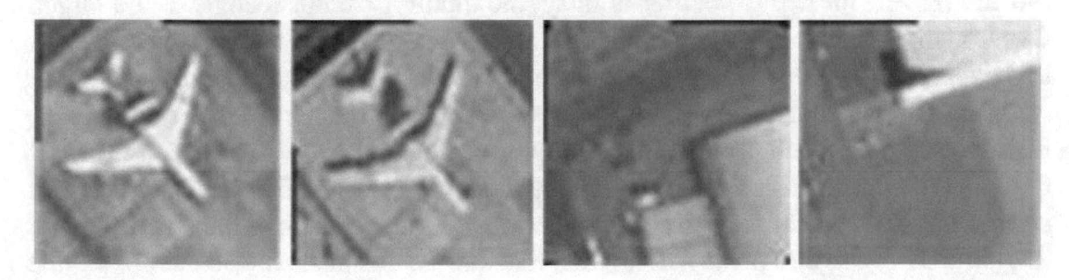

Fig. 3 Positive patches and negative patches

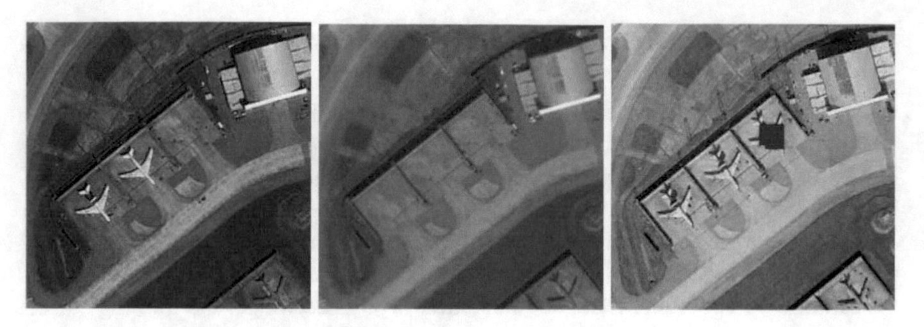

Fig. 4 Visualization of the monitoring results

on mnist dataset and get the similar result. So it is not likely overfitting. We test the classifier on the test image and it can give the right answer. We test many times. Our monitor can mark the image with airplane correctly. Figure 4 shows a success one. First two doesn't have airplanes in the target area, the last one have an airplane.

We even take one image as the training dataset, but this time the monitor makes two mistakes. It doesn't report two images which contains planes. We also test k = 2, the classifier makes one mistake this time. If the classifier remembers not enough, the monitor will not work. DCNN has a good memory. So far, it cannot tell different kinds of fighter plane because the classifier is a binary classifier.

These additional experiments demonstrate that we have to provide sufficient data.

3.3 Abnormal Weather Detection Scenarios

Beijing suffers from hazy weather in recent. The second dataset is from Beijing, China. It has more frames (Fig. 5 shows the fusion of the satellite image series of this area and part of the label layer). We can see many different shooting conditions. We take a building as reference because it changes very little. The weather conditions in the k images are clear and good. The image size is 128×128, so we have to do subsampling before we give it to the classifier.

Figure 6 shows the result given by the monitor. We can see most of time it can tell the target. But if the weather is hazy, the monitor cannot recognize the target. Because the training dataset do not have one hazy day. We can use this phenomenon to do abnormal weather detection if we choose normal weather as the training dataset.

More samples, more training time. So we also test different size (less than 2000) positive samples. With 400 positive samples, it makes 12 mistakes in ten times. It didn't tell the target in the image with many noise twice. The other is due to the mild contamination weather (row 2 col 2). The average time (generate dataset, train and test) is 133 s. With 800 positive samples, it makes 10 mistakes. It also cannot tell the target in mild contamination weather. The average time is 162 s. With 1200 positive samples, it makes 8 mistakes. It can tell the target in mild contamination weather

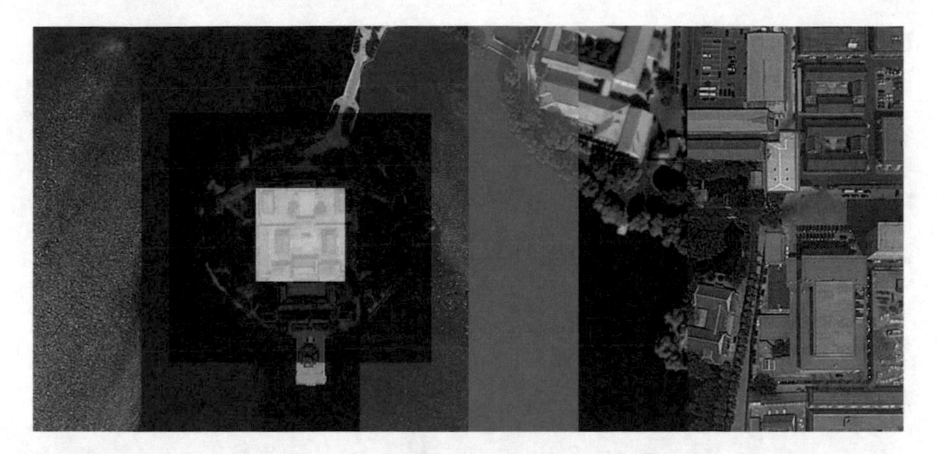

Fig. 5 The fusion of satellite image series and part of the label image

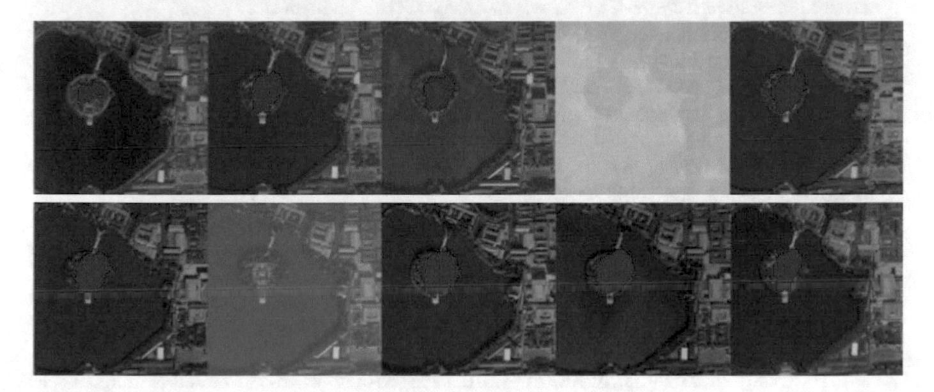

Fig. 6 Visualization of the monitoring results

twice in ten times. The average time is 184 s. All the configuration cannot tell the target in severe haze day.

We put the high resolution images and the source code in our Github repository. Now they are available at https://github.com/suixudongi8/SITS.

4 Conclusion

In this paper, we propose a method based on DCNN to monitor target in satellite image series. DCNN can learn features from raw data and is invariant to some noise like illumination change, small rotation and shift (good memory). The experiment shows that our method is effective in typical target monitoring tasks if the data is enough. The method can be used for reporting the activity of the fighter in military intelligence analysis scenarios and detecting abnormal events (hazy weather), depending on which class you define and give to the system.

In our future works, we will be devoted to improve the performance of the method. Two classes cannot represent more attribute of the target, which leaves much to be desired. We will add more classes to do better job. Adding the application of multi-target monitoring is also in our schedule.

Acknowledgments The authors would like to thank Jianjun Zhang and Yifei Fan for their constructive discussions and comments.

References

1. Li, Xi, & Li, Deren. (2014). Can night-time light images play a role in evaluating the syrian crisis? *International Journal of Remote Sensing*, *35*(18), 6648–6661.
2. LeCun, Yann, Bottou, Léon, Bengio, Yoshua, & Haffner, Patrick. (1998). Gradient-based learning applied to document recognition. *Proceedings of the IEEE*, *86*(11), 2278–2324.
3. Krizhevsky, A., Sutskever, I., & Hinton, G. E. (2012). Imagenet classification with deep convolutional neural networks. In *Advances in neural information processing systems* (pp. 1097–1105).
4. Simonyan, K., & Zisserman, A. (2014). Very deep convolutional networks for large-scale image recognition. arXiv:1409.1556.
5. Szegedy, C., Liu, W., Jia, Y., Sermanet, P., Reed, R., Anguelov, D., Erhan, D., Vanhoucke, V., & Rabinovich, A. (2015). Going deeper with convolutions. *Proceedings of the IEEE Conference on Computer Vision and Pattern Recognition* (pp. 1–9).
6. Salberg, A. B. (2015). Detection of seals in remote sensing images using features extracted from deep convolutional neural networks. *2015 IEEE International Geoscience and Remote Sensing Symposium (IGARSS)* (pp. 1893–1896).
7. Wu, H., Zhang, H., Zhang, J., & Xu, XU. (2015). Fast aircraft detection in satellite images based on convolutional neural networks. 2015 IEEE International Conference on Image Processing (ICIP) (pp. 4210–4214).
8. Hahn, M., Chen, S., & Dehghan, A. (2015). Deep tracking: Visual tracking using deep convolutional networks. arXiv:1512.03993.
9. CUDA Nvidia. (2007). Compute unified device architecture programming guide.
10. Chollet, F. (2015). Keras: Theano-based deep learning library. *Code*: https://github.com/fchollet. *Documentation*: http://keras.io.

A Method for Extracting Lexicon for Sentiment Analysis Based on Morphological Sentence Patterns

Youngsub Han, Yanggon Kim and Ikhyeon Jang

Abstract In these days, people share their emotions, opinions, and experiences of products or services using online review services on their comments, and the people concern the reviews to make decision when buying products or services. Sentiment analysis is one of the solution to observe and summarize emotional opinions from the data. In spite of high demands for developing sentiment analysis, the development of the sentiment analysis faces some challenges to analyze the data, because the data is unstructured, unlabeled, and noisy. The aspect-based sentiment analysis approach helps for more in-depth analysis, however building aspect and emotional expression is one of the challenge for the aspect-based sentiment analysis approach. Accordingly, we propose an unsupervised system for building aspect-expressions to minimize human-coding efforts. The proposed method uses morphological sentence patterns through an aspect-expression pattern recognizer. It guarantees relatively higher accuracy. As well as, we found some characteristics for selecting patterns to extracting aspect-expressions accurately. The greatest advantage of our system is performing without any human coded train-set.

Keywords Data mining · Lexicon building · Sentiment analysis · Aspect-based sentiment analysis

---

Y. Han (✉) · Y. Kim (✉)
Department of Computer and Information Sciences, Towson University,
Towson, MD 21204, USA
e-mail: yhan3@students.towson.edu

I. Jang (✉)
Department of Information and Communication Engineering,
Dongguk University Gyeongju, Gyeongbuk, South Korea
e-mail: ihjang@dongguk.ac.kr

R. Lee (ed.), *Software Engineering Research, Management and Applications*, Studies in Computational Intelligence 654,
DOI 10.1007/978-3-319-33903-0_7

1 Introduction

In these days, people share and post reviews of products or services with comments. The user-generated online reviews contain user's emotional state and opinion about topics or issues, such as events, products, entertainers, politicians, movies, services and so on [1, 2]. Furthermore, the people concern the reviews to make decision when buying products or services. As the increasing the population of the data, demanding of analyzing the data is increasing continuously [3]. Sentiment analysis is aimed to observe and summarize their sentimental opinions from the data. In spite of high demands for developing sentiment analysis, it faces some challenges because the data is unstructured, unlabeled, and noisy. In our previous research, we already proposed a probability model based sentiment analyzer [4]. It guarantees higher accuracy (89 %) than other approaches, and it can be broadly used to analyze text based data [5]. However, this model has some limitations to maintain the accuracy. The one of problems from the method is that all words have polarity while the words not containing meaningful information. For examples, determiners "a" or "the", and prepositions "to" or "on" have certain polarity. It causes incorrect results or over-analysis. Also, the approach requires human coded trains-set. It means the train-set must be re-built continuously to maintain the accuracy [4]. To minimize the limitations, we decide to apply the aspect-based sentiment analysis approach because it helps for more in-depth analysis [3, 6, 7]. For example, if we analyze movie reviews, some keywords are considered to more meaningful aspects such as "character", "story-line", "plot", "effect" and "music", and some keywords are considered to more meaningful emotional expressions as such as "good", "awesome", "amazing", "bad", "awful". However, building lexicon of aspect and emotional expression is one of the biggest challenge for the aspect-based sentiment analysis approach. Accordingly, we propose an unsupervised method for building aspect-expressions lexicons for aspect-based sentiment using morphological sentence. Thus, the main purpose of this system is to minimize human-coding efforts to building sentiment lexicon.

2 Related Works

2.1 Natural Language Processing

To analyze text based data, we used a natural language processing tool which is the "Stanford Core NLP" made by The Stanford Natural Language Processing Group. This tool provides refined and sophisticated results from text data based on English grammar such as the base forms of words, the parts of speech (POS), the named entity and the structure of sentences in terms of phrases and word dependencies [8]. In this research, we used the tool for two main reasons. The first reason is that the online textual data has a lot of linguistic problems to be analyzed such as spacing errors, idioms, and jargons. Another reason is that our method extracts morphological patterns to build lexicon of aspects and emotional expressions for further analysis using the structure of sentences and part of speeches.

2.2 Sentiment Analysis

The purpose of sentiment analysis is extracting opinion or emotional states regarding certain topics such as events, products, entertainers, politicians, movies, services from the text based data to find people's interesting and thought [5]. In this section, we will discuss about three existing approaches which are the "Lexicon based Sentiment Analysis", the "Probability Model based Sentiment Analysis", and the "Aspect based Sentiment Analysis".

2.2.1 Lexicon Based Sentiment Analysis

This approach is based on matching emotional words or word phrases using sentimental lexicon. Each word in the lexicon can be categorized as either positive or negative keywords. A message can be categorized as positive depending on the ratio of positive words versus negative words containing in the message. The accuracy of the existing approaches is as high as 80 % [4]. The results shows that the approach can be used as a supplement for traditional survey. However, the lexicon based approach has a weakness that even if a message contains positive words, the message doesn't necessary to categorize into positive. For example, the word "like" is categorized as a positive word in the lexicon because the word is usually used to express positive meaning but the word has different meaning such as "similar". If the message containing the word "like" must be categorized into positive, it is considered to error in terms of sentiment analysis. In this sense, such lexicon based approach should be improved regarding the nature of language.

2.2.2 Probability Model Based Sentiment Analysis

Lee et al. [4] proposed a method for sentiment analysis using the probability model. The method reads sample text messages in a train set and builds a sentiment lexicon that contains the list of words that appeared in the text messages and probability that a text message is positive opinion if it includes those words. Then, it computes the positivity score of text messages in a test set using the list of words in a message and sentiment lexicon. Each message is categorized as either positive or negative, depending on threshold value calculated using a train set. However, this model has some limitations and problems. For example, all of terms have positivity even though the term has not meaningful opinion such as "a", "the", "to" and so on. It causes over-analysis to extract opinion from the data. Also, the train-set must be up-to-date because the online textual data is trendy but the approach didn't concern it. This model seems a powerful approach but it need to improve in terms of the reasons.

2.2.3 Aspect Based Sentiment Analysis

The aspect based sentiment analysis is the lexicon based sentiment analysis because this approach also uses the lexicon as a measurement. However, this approach performs more in-depth sentiment analysis. In the aspect-based opinion mining, all result are categorized into each aspect which means features of entities or objects paired with expression. For example, if an object is a mobile phone, its aspects are the display, size, price, camera, or battery. In this case, aspects seems the attributes of the objects to describe more detail. Thus, expected result are "display-clean", "price-good", or "camera-awesome" from this approach while "clean", "good", or "awesome" from the lexicon based sentiment analysis [3, 6, 7]. Therefore, we decide to apply the aspect-based sentiment analysis approach for more in-depth analysis than our previous approach.

2.3 Lexicon Building

In sentiment analysis, the building lexicon is a fundamental challenge because the lexicon is a main measurement for extracting opinions from the data. For example, incorrect words or a less amount of words may cause a bad influence on the accuracy of result. Many studies have proposed unsupervised or semi-supervised approach for building lexicon. They were focused on constructing emotional lexicons that assign into fine-grained categories of emotions, such as happiness, like, disgust, sadness, and anger, have emerged recently [9–13]. Particularly, J. Bross, and H. Ehrig proposed a method that allows to automatically adapt and extend existing lexicons to a specific product domain, but they simply used morphological patterns to extract aspect-expressions [3]. Therefore, we proposes a system to extract aspects-expression lexicon using morphological pattern with focusing on how the patterns works effectively and accurately.

3 System Architecture and Implementation

The system consists of three main phases; collecting data, recognizing morphological sentence patterns, and extracting aspect-expressions. The first phase is data collection. The crawler collects movie review data from the IMDB, Rotten Tomatoes, and Metacritic. We developed crawler using a HTML parser. The second phase is morphological sentence pattern recognition. In this phase, the pattern

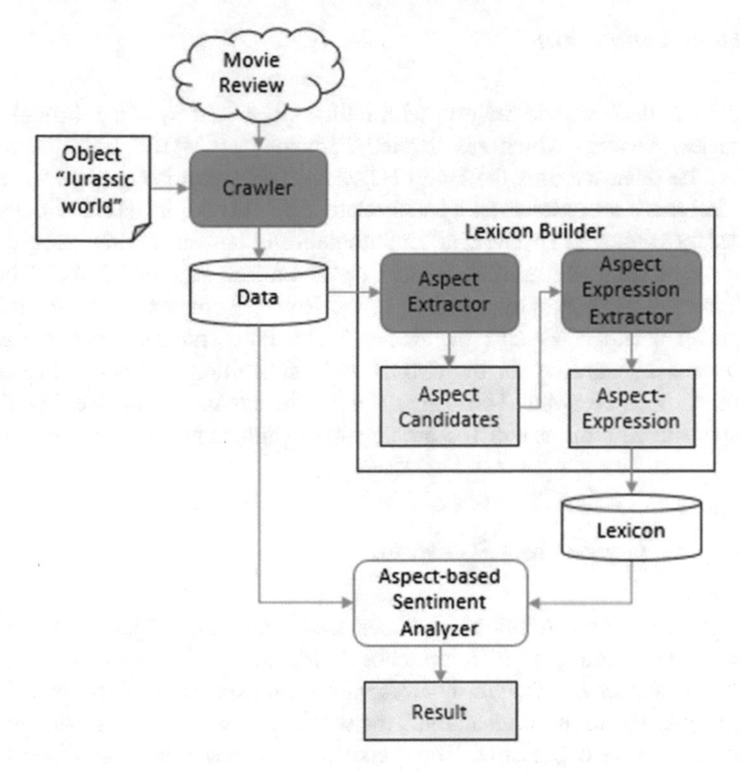

Fig. 1 The system architecture and flow

recognizer extracts morphological sentence patterns using aspect and expression candidates which are extracted through term extractor. At the last phase, the system extracts aspect-expressions using the patterns. Then, we proposed a manner of applying thresholds for the accuracy of extracting aspect-expressions. We expect that the aspect-expression lexicon can be used to aspect based sentiment analysis. The Fig. 1 shows architecture and flow of our proposed system.

The system consists of three main phases; collecting data, recognizing morphological sentence patterns, and extracting aspect-expressions. The first phase is data collection. The crawler collects movie review data from the IMDB, Rotten Tomatoes, and Metacritic. We developed crawler using a HTML parser. The second phase is morphological sentence pattern recognition. In this phase, the pattern recognizer extracts morphological sentence patterns using aspect and expression candidates which are extracted through term extractor. At the last phase, the system extracts aspect-expressions using the patterns. Then, we proposed a manner of applying thresholds for the accuracy of extracting aspect-expressions. We expect that the aspect-expression lexicon can be used to aspect based sentiment analysis. The Fig. 1 shows architecture and flow of our proposed system.

3.1 Data Collection

The crawler collects movie reviews with ratings generated by users through online movie review services which are "Rotten Tomatoes", "IMDB", and "Metacritic". To collect the data, we uses the Jsoup HTML parser which is an open-source Java library designed to extract and manipulate data stored in HTML documents developed by Hedley [14]. The crawler automatically collects review using a movie name as a seed such as "Jurassic World" or "Avengers: Age of Ultron". There are some difference in their rating scales. In the Rotten Tomatoes, a writer indicates their opinion weather "Fresh", or "Rotten". The Fresh means a positive and the Rotten means a negative. In the IMDB and Metacritic, a writer indicates their opinion from 1 to 10 point. The bigger number means more positive. We decided 8–10 are positive opinions and 1–3 are negative opinions to calculate their polarity.

3.2 Aspect-Expression Extractor

People indicates their opinion in a sentence including objects, aspects, and expressions. For example, in a sentence "This movie is awesome", the word "movie" is an aspect and the word "awesome" is an expression. Also, in a sentence "the movie was too focused on action", the word "action" is an aspect and the word "too focused" is an expression. This perspective of analysis called aspect based sentiment analysis [3, 6].

To extract aspect-expressions, the first step is the natural language processing (NLP) using a tool which is Stanford Core NLP made by The Stanford Natural Language Processing Group. This software is an integrated suite of natural language processing tools for English. It provided refined and sophisticated results based on English grammar. In this paper, we used a part of speech tagger and a sentence parser. The second step is extracting terms. The noun or noun phrase is an important part of speech because the subject word names a person, place or thing and the verb identifies an action or a state of being. An object receives the action and usually follows the verb [6]. Thus, noun or noun phrase is to be a clue word for next step. The second step is morphological pattern recognition. To extract more aspect candidates, the system extract morphological sentence patterns when the clue word occurs in the sentence. After this step, the system extract aspect candidates using the patterns. The system extracts expressions as extracting aspect. In addition, using the scores of movie reviews generated from the movie review crawler, the system calculates polarity (positive/negative), called the positivity in this paper, for sentiment analysis. The Fig. 2 shows the aspect candidate system flow.

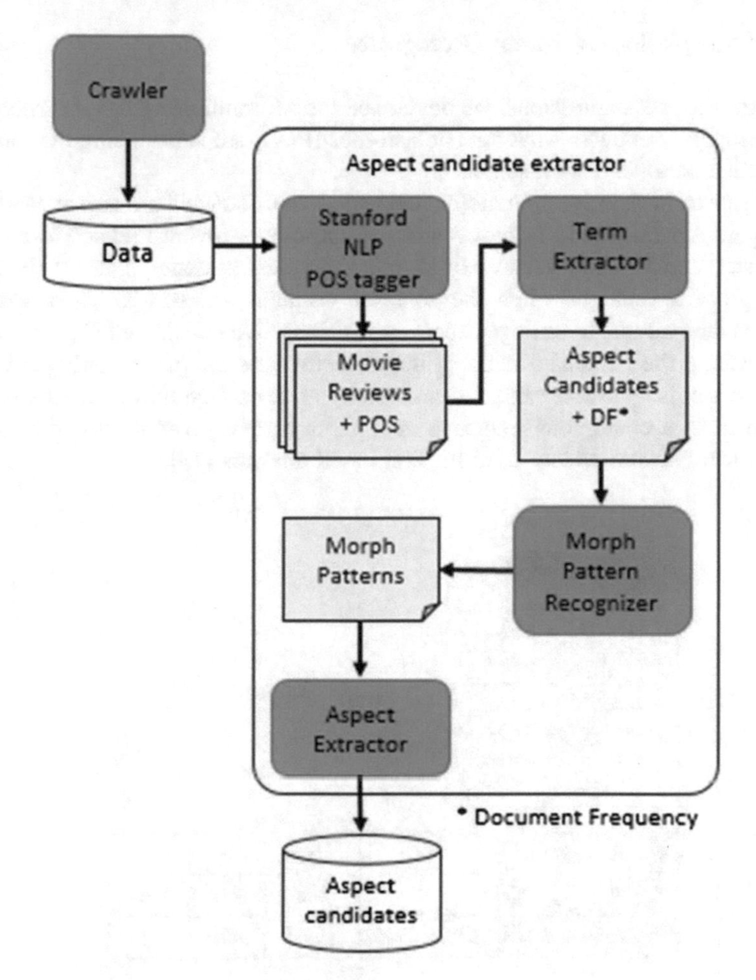

Fig. 2 Aspect candidate extractor system flow

3.2.1 Term Extractor

The term extractor extracts noun, noun phrase, pronoun, adjective, verb, and adverb as terms from the documents because the part-of-speeches may have meaningful opinions from the document in general [6]. The extracted terms can be used to discover important information related to the issues such as a product, company or movies from the documents. Also the extractor calculates document frequency (DF), is defined to be the number of documents in the collection that contain a term, to extract aspect candidates because it is used for extracting common used term from the collection of documents. And then, the system uses the top 100 terms which are most frequently used in the documents. The extractor stores part-of-speeches (POS) to extract further aspect-expression candidates.

3.2.2 Morphological Pattern Recognizer

To extract aspect-expressions, we developed the Morphological pattern recognizer. The recognizer extracts what Part of speeches (POS) are surrounding the aspect or expression candidates in a sentence.

As shown in the Fig. 3, the term "character" and "sequel" are aspect candidates which are top frequently occurred terms in the movie reviews related to a movie, "Jurassic World". The system extract morphological sentence patterns (POS patterns) from a sentence when the sentence contains the words. Then, extractor matches the patterns to retrieve aspect-expressions. We considered N-grams model for matching the patterns because if the extractor uses the pattern only perfect/full matched words as aspect candidates, diversity of extraction may decrease. N-gram is defined as a contiguous sequence of n items from a given sequence of text or speech and N-gram widely used for text based analysis [15].

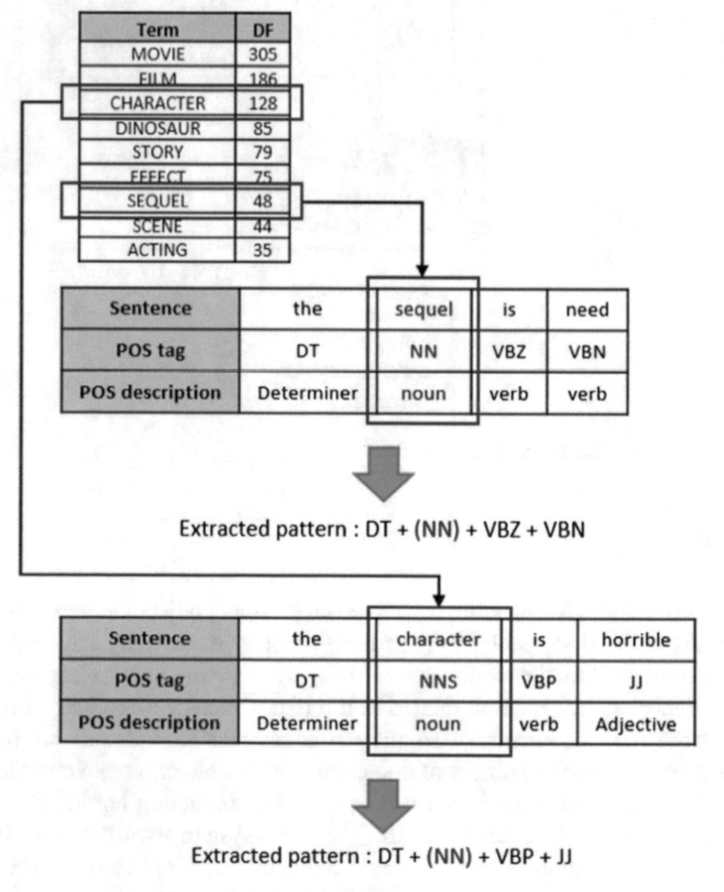

Fig. 3 Example of extracting morphological sentence patterns

Table 1 Examples of extracted morphological sentence pattern using terms. (P is POS, Number is Sequence of POS)

Len.	Prefix			Aspect	Postfix		
	**Pi* − 3*	*Pi* − 2	*Pi* − 1	*Wi*	*Pi* + 1	*Pi* + 2	*Pi* + 3
6	PRP$	JJR	IN	JJ NNP	CD	,	CC
5	PRP$	JJR	IN	JJ NNP	CD	,	
5		JJR	IN	JJ NNP	CD	,	CC
4		JJR	IN	JJ NNP	CD	,	
3		JJR	IN	JJ NNP	CD		
3			IN	JJ NNP	CD	,	
2			IN	JJ NNP	CD		

**P-n* Sequence of POSs

The Table 1 shows an example of the extracted patterns after applying n-gram model. At the first row in the table, the pattern consists of 6 length sequence of POSs (PRP$ + JJR + IN + JJ + NNP + CD+, +CC). The aspect and expression extractor matches all possible patterns. In addition, the longest pattern has a more priority to avoid duplicate extraction.

After Morphological pattern reorganization, aspect extractor retrieves aspects when the morphological patterns are matched and the aspect candidates is are occurred in a sentence. The Table 2 shows examples of extracted aspect candidates. The total number of extracted words are 4,999 words through the extract using all patterns while 450 words through the term extractor as aspect candidates from same 1,000 movie reviews. It implies that our aspect candidate extractor could find the more number of candidates than the term extractor.

3.2.3 Aspect Extractor

After Morphological pattern reorganization, aspect extractor retrieves aspects when the morphological patterns are matched and the aspect candidates is are occurred in

Table 2 Examples of extracted aspect candidates from aspect candidate extractor

Rank	Expression	Count
1	MOVIE	882
2	FILM	678
3	DINOSAUR	498
4	CHARACTER	403
5	JURASSIC PARK	367
6	ORIGINAL	362
7	JURASSIC WORLD	222
8	ACTION	174
9	STORY	173
10	SEQUEL	168

Table 3 Examples of extracted expression candidates from aspect candidate extractor

Rank	Expression	Count
1	GOOD	282
2	BETTER	155
3	FUN	149
4	ENOUGH	136
5	WELL	104
6	BLOCKBUSTER	99
7	STUPID	94
8	BEST	93
9	BIG	81
10	BAD	80

a sentence. The Table 2 shows examples of extracted aspect candidates. The total number of extracted words are 4,999 words through the extract using all patterns while 450 words through the term extractor as aspect candidates from same 1,000 movie reviews. It implies that our aspect candidate extractor could find the more number of candidates than the term extractor.

3.2.4 Expression Extractor

Expression extractor retrieves expressions as the aspect candidate extractor. The Table 3 shows examples of extracted aspect candidates. The total number of extracted words are 5,096 words through the extract using all patterns while 450 words through the term extractor as expressions. It show similar results as aspect candidate extractor.

4 Experiment Result

This section presents the experiment results of lexicon building method we proposed. We collected 1,000 reviews related to a Movie, "Jurassic world" using the data crawler. Then, we selected 1,000 sentences from the data for experiments. To evaluate our method, we built aspects and expression lexicon as answer-sets which is labeled by human annotators. We found 230 aspects and 250 expressions in this study.

4.1 Measurement

To evaluate the performance, we calculated the F-measure which is broadly used to measure the performance for this type of systems [16]. Also, we compered the measures with existing related researches at the end of this section. The definition of the measure is:

$$\text{Recall} = \frac{\text{TP}}{\text{TP} + \text{FN}} \tag{1}$$

$$\text{Precision} = \frac{\text{TP}}{\text{TP} + \text{FP}} \tag{2}$$

$$\text{F} - \text{measure} = 2 * \frac{\text{Precision} * \text{Recall}}{\text{Precision} + \text{Recall}} \tag{3}$$

The F-measure considers the "Recall" and "Precision". The recall means the portion of relevant instances that are retrieved, and the precision means the portion of retrieved instances that are relevant. In this equations, TP is true positive, TN is true negative, FP is false positive, and FN is false negative. For example, when an extracted aspect or expression is classified as 'correct' by the extractor while the aspect or expression is labelled by answer-set which is labeled by human annotators is 'related', the aspect or expression is considered TP (true positive). On the other hand, when an extracted aspect or expression is classified as 'correct' by the extractor while the aspect or expression is labelled by answer-set which is labeled by human annotators is 'non-related', then the aspect or expression is considered TN (true negative).

4.2 Morphological Sentence Pattern Selection

4.2.1 Extracting Aspect

The recognizer generates the morphological sentence patterns using the N-gram method from original patterns as seeds which are containing top 100 aspects extracted through term extractor. The recognizer generated 59,111 patterns (2–20 lengths patterns) from 1,000 sentences of movie reviews. To use the patterns to extract aspect candidates, it is required to verify what types of patterns can extract aspect candidates effectively and accurately. Accordingly, we examined what type of patterns can extract the most number of correct in terms of the length and frequency. In the Fig. 4, 3–7 lengths patterns could extract 90.29 % (1,833 out of 2,030). Therefore, we select the 3–7 lengths patterns to use for extracting aspects.

At this point, we examined to compare performance with "Noun Phrase", "Aspect Pattern", "Aspect Sentence Pattern", and "3–7 Sentence pattern" using the F-measure. The "Noun Phrase means" that using only noun phrase to extract aspect because most of aspects consist of noun based phrase [6, 7]. The Aspect Pattern means that we extract the patterns based on our aspect candidates extracted through term extractor. The Aspect Sentence Pattern means that we extract the patterns

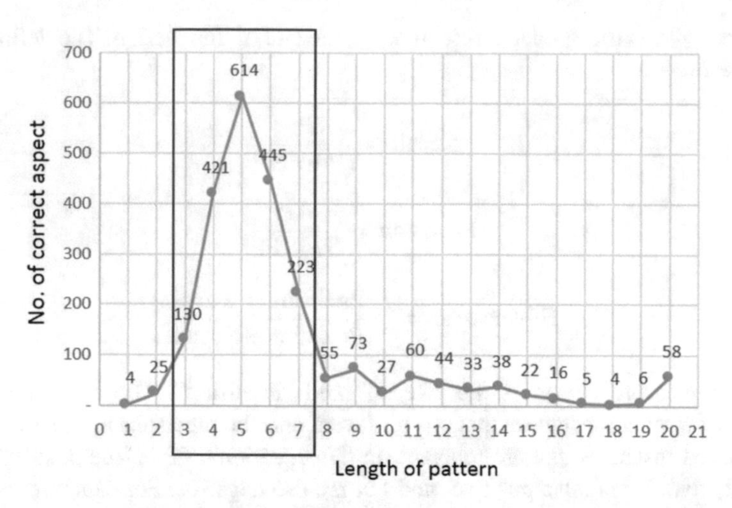

Fig. 4 The numbers of extracted correct aspect by the length of patterns

extracted through our pattern recognizer. The 3–7 Aspect Sentence pattern means that we selected 3–7 lengths patterns from all extracted patterns through the pattern recognizer. The Table 4 show the results of precisions, recalls and F-scores for each type of patterns. As shown in the table, the 3–7 patterns can extracted more accurately (F-score 80.86 %). Thus, we decided to use 3–7 lengths patterns.

4.2.2 Extracting Expression

As the experiment of extracting aspect, the recognizer generates 21,101 patterns (2–20 lengths patterns) from 1,000 sentences of movie reviews. We verified patterns in the same manner of the extracting aspect. In the Fig. 5, the 2–6 lengths

Table 4 The F-score of extracted aspect by the types of patterns

Pattern	Extracted	Correct	Answer	Precision (%)	Recall (%)	F-score (%)
Noun phrase	450	90	230	20.00	39.13	26.47
Aspect pattern	4,999	230	230	4.60	100.00	8.80
Aspect sentence pattern	1,068	227	230	21.25	98.70	34.98
3–7 sentence pattern	329	226	230	68.69	98.26	80.86
3–7 sentence pattern (freq. >= 1)	236	205	230	86.86	89.13	87.98
Co-occurrence	243	208	230	85.60	90.43	87.95

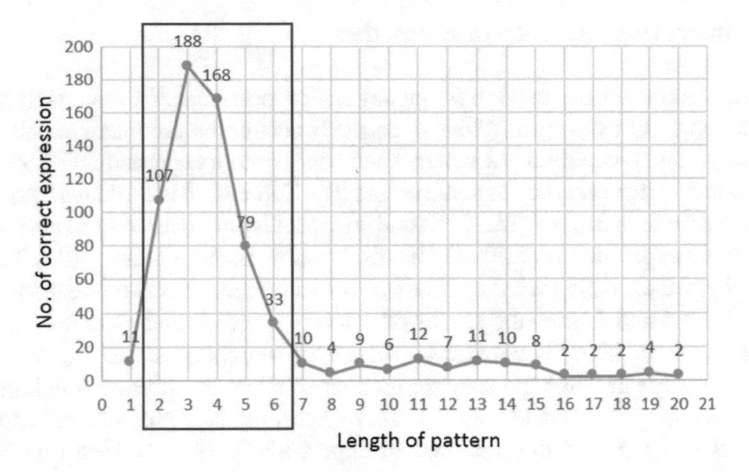

Fig. 5 The numbers of extracted correct expression by the length of patterns

patterns could extract 90.29 % (1,833 out of 2,030). Therefore, we select the 2–6 lengths patterns to use for extracting expressions.

At this point, we examined to compare performance using the F-measure with "Adj. Phrase", "Expression Pattern", "Expression Sentence Pattern", and "2–6 Sentence Pattern". The Adj. Phrase means that using adjective phrase because most of expression consist of adjective based phrase [6, 7].

The Expr. Pattern means that we extract the patterns based on our expression candidates extracted through term extractor. The Expr. Sentence Pattern means that we extract the patterns extracted through our pattern recognizer. The 2–6 Sentence Pattern means that we selected 2–6 lengths patterns from all patterns extracted through our pattern recognizer. The Table 5 show the results of precisions, recalls and F-scores for each type of patterns. As shown in the Table V, the 2–6 Sentence Pattern can extract more accurately (F-score 58.01 %). Thus, we decided to use 2–6 lengths.

Table 5 The F-score of extracted expression by the types of patterns

Pattern	Extracted	Correct	Answer	Precision (%)	Recall (%)	F-score (%)
Adj. phrase	450	54	250	12.00	21.60	15.43
Expression pattern	5,096	250	250	4.91	100.00	9.35
Expression sentence pattern	610	202	250	33.11	80.80	46.98
2–6 sentence pattern	605	248	250	40.99	99.20	58.01
2–6 sentence pattern (freq. >= 1)	329	220	250	66.87	88.00	75.99
Co-occurrence	283	206	250	72.79	82.40	77.30

4.2.3 Improving Accuracy of Extraction

We found characteristics that the frequency and co-occurrence of aspect-expressions affect accuracy of extraction. When an aspect is occurred more than once (87.98 %) and an aspect is co-occurred in a sentence with one or more expressions (87.95 %), the results show higher accuracy than the others (See Table 4). Also, when an expression is occurred more than once (75.99 %) and an aspect is co-occurred in a sentence with one or more expressions (77.30 %), the results show higher accuracy (See Table 5). Especially, we examined what numbers of co-occurrence affects to the result.

The Figs. 6 and 7 show the results of precisions, recalls and F-scores depending on the numbers of co-occurrences. When the numbers of an aspect and an expression is greater than the average number of all co-occurrence which numbers are 227 for the aspects and 129 for the expressions, the precision of aspects is 100 % (See Fig. 6) and the precision of expression is 99.14 % (See Fig. 7). This finding suggests.

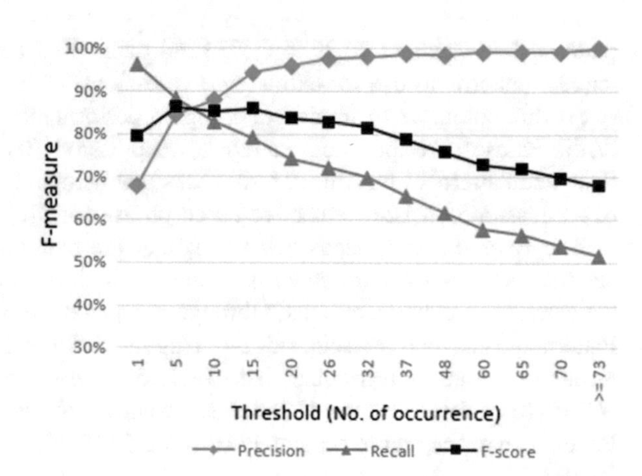

Fig. 6 The F-score of extracted correct aspect by the length of patterns

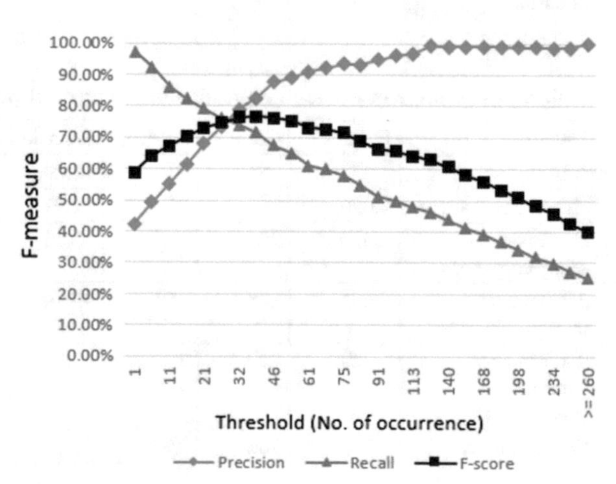

Fig. 7 The F-score of extracted correct expression by the length of patterns

Table 6 Comparison of F-score with related researches

Methods	F-score
HL [17]	60.49
MPQA [18]	59.15
NRC-Emotion [20]	54.81
HashtagLex [19]	65.30
Sentiment140Lex [19]	72.51
TS-Lex [16]	78.07
Proposed system	**82.81**

4.2.4 Comparison with Related Researches

Table 6 shows results of F-measures for comparison with related researches. We compare our result with HL [17], MPQA9 [18], HashtagLex and Sentiment140Lex11 [19], and TS-Lex [16]. HL, MPQA and NRC-Emotion [20] are traditional model with a relative small lexicon. HashtagLex, Sentiment140Lex and TS-Lex are sentiment lexicons for Twitter. F1 score of our approach is 82.81 which is relatively higher than other approaches.

5 Conclusion

This research proposed a method for building aspect-expression lexicon for aspect-based sentiment analysis using morphological sentence patterns that guarantees relatively higher accuracy (F-measure) than existing approaches. Through the results, we found some characteristics for selecting patterns to extract aspect-expressions accurately. The first characteristic is the length of pattern. Therefore, we suggest 3–7 lengths pattern for extracting aspects and 2–6 lengths patterns for extracting expressions. The second characteristic is the frequency of extracted aspect-expressions. More frequently occurred aspect-expression tend to more accurate. The last characteristic is co-occurrence of aspects and expressions. The more number of co-occurrences tend to more accurate. We can assume that aspects can be clues for extracting expression and expressions can be clues for extracting aspects. Also, we suggest that the threshold can be adjusted depending on needs. For example, when a user concerns the higher precision while the recall, the user adjust threshold to be higher following our results. Finally, our system shows a exceeding performance compared with existing emotional lexicon building approaches in terms of accuracy, and it performs without any human coded train-set and knowledge-base. The future work is how our system performs to cross domain such as social media (Tweets, YouTube comments), News and so on. We can assume that the data has different characteristic to extracting information from that sources. Also, we will apply this method for improving the probability model based sentiment analysis.

References

1. Sharma, A., & Dey, S. (2012). A comparative study of feature selection and machine learing techniques for sentiment analysis. *Proceedings of the 2012 ACM Research in Applied Computation Symposium* (pp. 1–7), October 2012. ISBN: 978- 1-4503-1492-3.
2. Goncalves, P., Araújo, M., Benevenuto, F., & Cha, M. (2013). Comparing and combining sentiment analysis methods. *Proceedings of the First ACM Conference on Online Social Networks* (pp. 27–38), October 2013. ISBN: 978-1-4503-2084-9.
3. Bross, J., & Ehrig, H. (2013) *Automatic Construction of Domain and Aspect Specific Sentiment Lexicons for Customer Review Mining, CIKM'13*, October 27–November 1. 2013. San Francisco, CA, USA, ISBN: 978-1-4503-2263-8.
4. Lee, H., Han, Y., Kim, Y., & Kim, K. (2014). Sentiment analysis on online social network using probability Model. *Proceedings of the Sixth International Conference on Advances in Future Internet* (pp. 14–19).
5. Melville, P., Gryc, W., & Lawrence, R. D. (2009). Sentiment analysis of blogs by combining lexical knowledge with text classification. *Proceedings of the 15th ACM SIGKDD International Conference on Knowledge Discovery and Data Mining* (pp. 1275–1284), June 2009. ISBN: 978-1-60558-495-9.
6. Thet, T. T., Na, J.-C., & Khoo, C. S. G. (2010). Aspect-based sentiment analysis of movie reviews on discussion boards. *Journal of Information Science, 36*, 823–848.
7. Wogenstein, F., Drescher, J., Reinel, D., Rill, S., & Scheidt, J. (2013). *Evaluation of an Algorithm for Aspect-Based Opinion Mining Using a Lexicon-Based Approach, WISDOM'13*, August 11. Chicago, USA. ISBN: 978-1-4503-2332-1.
8. Manning, C.D., Surdeanu, M., Bauer, J., Finkel, J., Bethard, S.J., & McClosky, D. (2014). The stanford CoreNLP natural language processing toolkit. *Proceedings of the 52nd Annual Meeting of the Association for Computational Linguistics: System Demonstrations* (pp. 55–60), June 2014.
9. Kaji, N., & Kitsuregawa, M.: Building lexicon for sentiment analysis from massive collection of html documents. *Proceedings of EMNLP-CoNLL* (pp 1075–1083), 2007.
10. Xu, J., Xu, R., Zheng, Y., Lu, Q., Wong, K.-F., & Wang, X. (2013). Chinese emotion lexicon developing via multi-lingual lexical resources integration. *Proceedings of the 14th International Conference on Computational Linguistics and Intelligent Text Processing* (Vol. 2, pp. 174–182).
11. Xu, L., Lin, H., Pan, Y., Ren, H., & Chen, J. (2008) Constructing the afective lexicon ontology. *Journal of the China Society For Scientific and Technical Information, 27* (2):180–185
12. Zhang, Z., & Singh, M. P. (2014) Renew: A semi-supervised framework for generating domain-specific lexicons and sentiment analysis. *Proceedings of the 52nd Annual Meeting of the Association for Computational Linguistics* (pp. 542–551).
13. Tai, Y.-J., & Kao, H.-Y. (2013). Automatic Domain-Specific Sentiment Lexicon Generation with Label Propagation. *IIWAS '13 Proceedings of International Conference on Information Integration and Web-based Applications and Services* (p. 53). ISBN: 978-1-4503-2113-6.
14. Hedley, J. Jsoup HTML parser. http://jsoup.org/.
15. Tomovic, A., Janicic, P., & Keselj, V. (2006). n-Gram-based classification and unsupervised hierarchical clustering of genome sequences. *Journal of computer methods and programs in biomedicine, 81*, 137–153.
16. Tang, D., Wei, F., Qin, B., & Zhou, M., Liu, T. (2014). Building large-scale twitter-specific sentiment lexicon. *Proceedings of the 25th International Conference on Computational Linguistics: Technical Papers* (pp. 172–182). Dublin, Ireland.
17. Hu, M., & Liu, B. (2004). Mining and summarizing customer reviews. *Proceedings of the ACM SIGKDD International Conference on Knowledge Discovery & Data Mining (KDD-2004, full paper)*, Aug 22–25. Seattle, Washington, USA.

18. Wilson, T., Wiebe, J., & Hoffmann, P. (2005). Recognizing contextual polarity in phrase-level sentiment analysis. *Proceedings of HLT-EMNLP-2005* (pp. 347–354). Stroudsburg, PA, USA. doi:10.3115/1220575.1220619.
19. Mohammad, S. M., Kiritchenko, S., & Zhu, X. NRC-Canada: Building the State-of-the-Art in sentiment analysis of tweets. *Proceedings of the International Workshop on Semantic Evaluation*.
20. Mohammad, S. M., & Turney, Peter D. (2012). Crowdsourcing a word–emotion association lexicon. *Computational Intelligence, 29*(3), 436–465.

A Research for Finding Relationship Between Mass Media and Social Media Based on Agenda Setting Theory

Jinhyuck Choi, Youngsub Han and Yanggon Kim

Abstract We are living in a flood of information. We hear about lots of social issues such as politics and economies in every day from the mass media. Before the appearance social media, it is difficult to interact people's opinions with the others about the social issues. However, we can analyze important social issues using big data generated from social media. We tried to apply the relationship between agenda setting theory and social media because we have received social issues from official accounts like news using social media, and then users shared social issues to other users, so we choose tweets of Baltimore Riot to analyze. We collected tweets related with Baltimore Riot, and then we extracted term keywords using text mining technologies such as TF-IDF. Actually, we analyzed tweets of 04-27-2015 Based on detected important words, we analyzed tweets at 5-min intervals, and we extracted tweets of mass media and others. Based on user's profiles, we found relationship of mass media and social issues. About initial phase of the social issues as it happened, local mass media leaded about incidents, and tweets exchanged and shared in local area. After writing an influence Twitter user, social issues of Baltimore Riot spread to other areas. As a result, we could detect agenda setting theory in social media using big data technology. It implies that the local mass media led the social issues, the Baltimore Riot became one of major social issues to people at the end.

Keywords Agenda setting theory · Data mining · Baltimore riot · Keyword extraction

J. Choi (✉) · Y. Han (✉) · Y. Kim (✉)
Department of Computer and Information Sciences, Towson University,
Towson, MD 21204, USA
e-mail: jchoi16@students.towson.edu

Y. Han
e-mail: yhan3@students.towson.edu

Y. Kim
e-mail: ykim@towson.edu

R. Lee (ed.), *Software Engineering Research, Management and Applications*, Studies in Computational Intelligence 654,
DOI 10.1007/978-3-319-33903-0_8

1 Introduction

We have been living in the era of smartphones since Steve Jobs left his great legacy, IPhone. In recent years, we have experienced swift change in our communication style because of social media and smartphones. After the appearance of smartphones, people can share news, information, and even their emotions through the social media as smartphones enabled people to access to the Internet anywhere and anytime they want. Also, social media provide diverse platforms such as Twitter and Facebook, which have an incredible amount of influence in the recent society. For instance, Twitter has become one of the most famous social media in the world since 2006 [1]. Additionally, according to the statistics of Twitter Company, the number of its monthly active users is 302 M, and the percentage of active users on mobile such as a smartphone and tablet PC is 80 % [2]. All these statistics are the analysis materials of December 31, 2015 cited from the official website of Twitter. It means Twitter is currently one of the major social media, so we planned to analyze Twitter among social media platforms.

There are three reasons why we picked Twitter among a lot of social media. First of all, Twitter is one of the most influential social media in the world, so it is easy to collect a lot of data for analyzing unpredictable incidents such as Baltimore Riot. Second of all, a Twitter user can write comment up to 140 characters, so most of them write their comments by implication to express their message more effectively. Thus, with the comments full of implications, we can easily extract the core of the messages which the users tried to express. Third, as many users and media companies post a lot of news and opinions in real time, it is easy to collect meaningful information through Twitter. Recently, this kind of information is actively discussed and studied for researches about analyzing Twitter.

As we know, we are now living in the information age, and we are introduced to numerous social issues through the traditional mass media such as newspaper or broadcasting. However, we may not easily catch every social issue as we can be overwhelmed by too much information that the mass media gives us every day, such as issues about politics, economy, education, or entertainment. Since the appearance of the traditional forms of mass media like a newspaper or broadcasting, we have depended on these forms of mass media in our daily lives whenever we tried to get any kind of information about our society. It means that people were restricted only to the information or opinions that the mass media gave although the traditional mass media allowed people to easily access to the information that covers a wide variety of social issues including terrorism or illness. In other words, we had no choice to decide which information or social issue was important to us; we were forced to regard what the mass media emphasize as the important information in our society.

Before the appearance of social media, as mentioned above, along with the fact that the traditional mass media could not provide the all kinds of information or opinions, there is another weakness in the traditional mass media. Since the mass media offered offline services only, it was difficult for the public to catch a general consensus about the important social issues immediately. That is because, the traditional mass media required time for offline public opinion poll using questionnaires or telephone surveys to figure out the public opinion, and to analyze the social issues. Also, limitations of the scope in public opinion poll such as the small size of samples in surveys made it much harder for people to get the precise public opinion about the social issues. However, due to the appearance of social media, we can analyze social issues much easier though the new methods from big data than the traditional methods like offline questionnaires. It means that people can access to the important social issues and the various opinions about them immediately based on the technology of big data. Therefore, we planned to analyze the relationship between Twitter of social media and the traditional mass media. For our research, we used the methods including TF-IDF and filtering for accuracy.

In the media field, there is a theory which is agenda setting theory. Agenda setting theory is the ability to influence the salience of topics on the public agenda [3]. It means if mass media emphasize their news to people, we would realize this issues are important. Based on agenda setting theory, as case study, we chose Baltimore Riot because it is one of the unpredictable social issues. As a result, we found interesting things of relationship between social media and mass media. As whole tweets, Twitter users liked to communication with their opinions with Twitter. However, as splited times, in the beginning, tweets of local mass media, journalist, and reporter led a social issue as local issues of Baltimore Riot, and then an influential Twitter user posted a tweet of Baltimore Riot, so tweets of Baltimore Riot were spread as a major social issue. It means that it was one of the important social issues because of agenda setting theory of mass media.

For analyzing tweets related to the Baltimore Riot, we constructed the Twitter Analyzer System. The system has 2 divided parts: crawler and analyzer. Crawler is a part for collecting tweets as seeds such as Baltimore or Baltimore Riot. Analyzer is a part for detecting important words. We calculated numerical values using the TF-IDF method at 5-min intervals, and we also searched user profiles of the Twitter users in order to figure out the relationship between the mass media and other Twitter users [4].

2 Related Work

2.1 Agenda Setting Theory

Agenda setting theory defines how mass media can be affected to people [5]. As traditional ways, we have used questionnaires and telephone surveys as manual sorting [6]. However, after the appearance social media such as Twitter, we could easily get what is important social issues because social media users like to post tweets in real time, so there are lots of tweets in online. Based on agenda setting theory, we wanted to catch relationship with mass media and tweets. Approaching to detect social issues from mass media in social media, we used tweets.

2.2 *Keyword Extraction*

Approaching agenda setting theory in social media, keyword extraction is definitely important because if we do not know important keywords, we could not realize what is important social issues, so we used TF-IDF to detect important keywords during that time [7].

2.2.1 Natural Language Processing (NLP)

We used a natural language processing tool to analyze text based data such as tweets. This tool is the "Stanford Core NLP" made by The Stanford Natural Language Processing Group [8]. Because of this tool, we could get refined and sophisticated results from text data. In this paper, we used Natural Language Processing to get refined data from tweets.

2.2.2 Term Frequency-Inverse Document Frequency (TF-IDF)

In this research, we used Term Frequency-Inverse Document Frequency) TF-IDF for the system to calculating numerical values [9]. Typically, TF-IDF is composed by two terms. First, it is Term Frequency which measures how frequently a term occurs in a document. It means that Term Frequency is the significance of a term in a document [10, 11]. Second is Inverse Document Frequency which measures the significance of a term in connection with the whole corpus [5, 12, 13]. Thus, the TF-IDF is often used in information retrieval and text mining. It means that we could get important meaningful words using TF-IDF.

The conventional TF-IDF weight scheme is defined as [10]:

$$tfidf_{i,j} = tf_{i,j} \times idf_i$$

TF value is raw frequency normalized by the number of times all terms appears in a document to prevent a bias towards longer documents. TF values is calculated as:

$$tf_{i,j} = \frac{n_{i,j}}{\sum_k n_{i,j}}$$

where $n_{i,j}$ is the number of times term i in document d_j and $\sum_k n_{i,j}$ is the number of time all terms appears in document d_j.

Also, IDF of a term can be defined as:

$$idf_i = log \frac{N}{|\{d \in D : t \in d\}|}$$

where $|\{d \in D: t \in d\}|$ is the number of documents in which term t appears and N is total number of all documents in a corpus.

In this research, a tweet is a document, so we could get numerical values as term frequency, and tweets at 5-min intervals are a document for Inverse Document Frequency.

3 System Architecture

Instead of using analysis of a traditional social issues such as questionnaire and telephone survey, this research used data-mining techniques to catch relationship with massive amount of tweets about Baltimore Riot using TF-IDF and the number of tweets (buzz). We applied to automated Twitter data-collecting tool to collect tweets for this research [14].

This research needs to build Twitter crawler and analyzer. For analyzing tweets, we collected the number of 1,967,451 tweets during 04-22-2015–05-12-2015 as seeds such as Baltimore and Baltimore Riot. We calculated TF-IDF from all collected tweets during at that time, and then we chose the number of 28,333 tweets during 04-27-2015 because this day was Baltimore riot's day, and then we extracted 5,998 user profiles including user id and description based on meaningful words from numerical numbers of TF-IDF.

3.1 Twitter Craweler

As shown in the Fig. 1, the Twitter crawler has 3 divided different parts which are Twitter Access, Real-time Search, and Duplication Check. First of all, to access Twitter, we need to get 4 keys such as ConsumerKey, ConsumerSecret, AccessToken, and AccessTokenSecret from Twitter. If all of keys are correct, we can access into Twitter. It means we can use Twitter data for collecting. Second of all, if we choose seeds such as 'Baltimore Riot', we would collect tweets including 'Baltimore Riot', but Twitter has limit of collecting tweets. It means a Twitter app key can give the system 180 quires during 15 min. To reduce time-wasting and inefficient, we made a total of 5 app keys to collect tweets. For example, if an app key finishes to collect data within 15 min, another twitter app key should be operated to collect tweets in real time. Third of all, the system have to check duplication check of tweets because duplicated tweets are the problem for accuracy. Especially, a tweet have a unique id consisted of numbers like 0421942. As a result, because of Twitter crawler, tweets would be collected as seeds for Twitter analyzer [15].

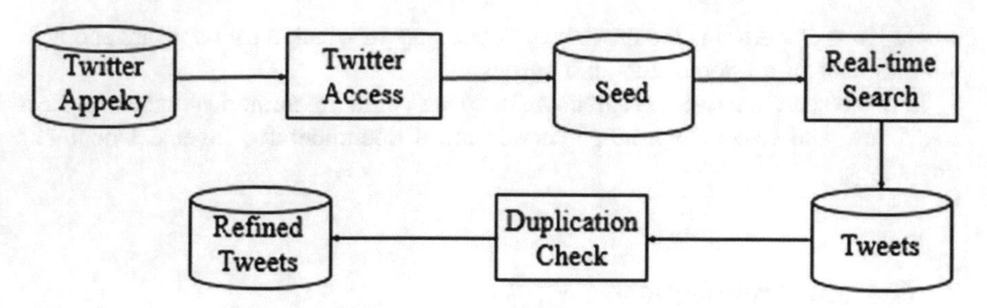

Fig. 1 The system architecture of Twitter crawler

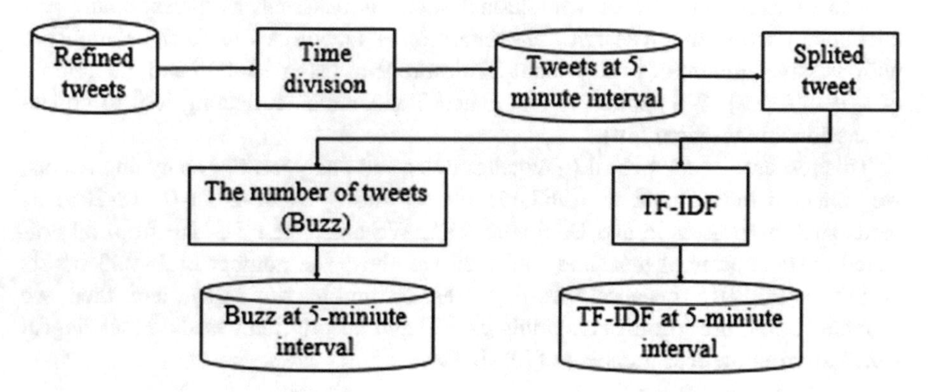

Fig. 2 The system architecture of Twitter analyzer

3.2 Twitter Analyzer

As shown in the Fig. 2, the Twitter analyzer has 4 divided different parts which are Time Division, Splited Tweet, The Number of Tweets(Buzz), and TF-IDF. First, based on collected tweets related with seeds, we recollected tweets at 5-min interval to catch important words. Second, based on recollected tweets, we extracted words to get numerical values of term frequency and TF-IDF, so we could get important words at 5-min interval. Third, based on extracted words, the system could get buzz, term frequency, and TF-IDF. To get numerical values to detect important words, we used all data during 04-22-2015–05-12-2015. Whole tweets are 1,967,451.

A tweet have 3 different categories; Singleton, Retweet, and Reply [1]. A Singleton is to write a tweet for oneself, so it is useful to find the first writer. If a Twitter user reposted a tweet by someone's tweets, it is Retweet. Retweet is marked by 'RT'. Tweets including 'RT' are useful to detect relationship of tweets of mass media and others. If a user writes a tweet by another user, we could see @id, it is reply. Reply is also useful to detect tweets of mass media and others based on user profiles. Based on collected tweets, we could get user's profiles such as user id and location. It is useful to classify tweets of mass media and others (Fig. 3).

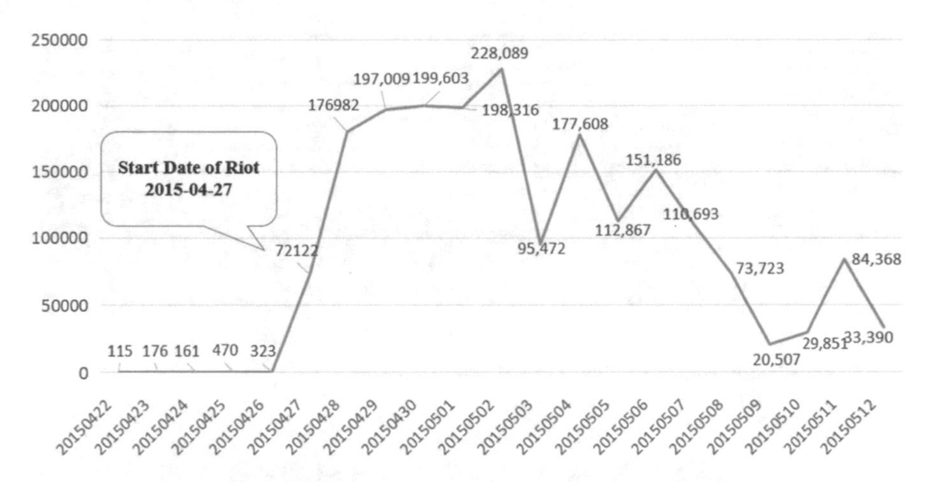

Fig. 3 Whole tweet buzz about Baltimore and Baltimore Riot

4 Result

We collected tweets related to Baltimore Riot as seeds such as Baltimore and Baltimore riot. As a result, during 04-22-2015 and 05-12-2015, whole tweets was 1,967,451. All data used for calculating meaningful words using TF-IDF. Especially, we analyzed tweets written by 04-27-2015 2:25 p.m.–7:30 p.m. because it was important time of Baltimore riot, total tweets were 28,333. The number of related tweets of mass media were 3,926, 14 %. The others were the number of 24,407 tweets, 86 % as shown in the Fig. 4. So, we assumed that Twitter users looked like to lead social issues in the online world as whole buzz (Fig. 5).

Fig. 4 Percentage of tweet buzz about mass media and others

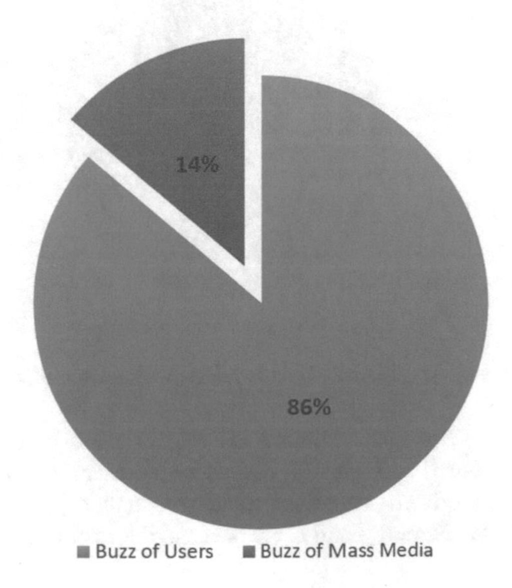

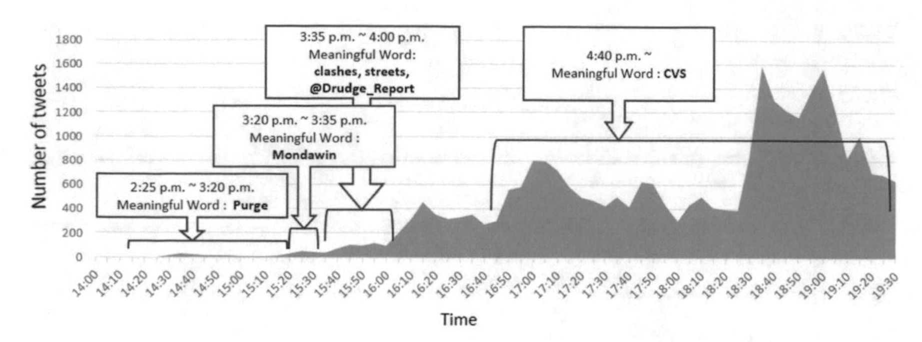

Fig. 5 Meaningful words using TF-IDF and term frequency

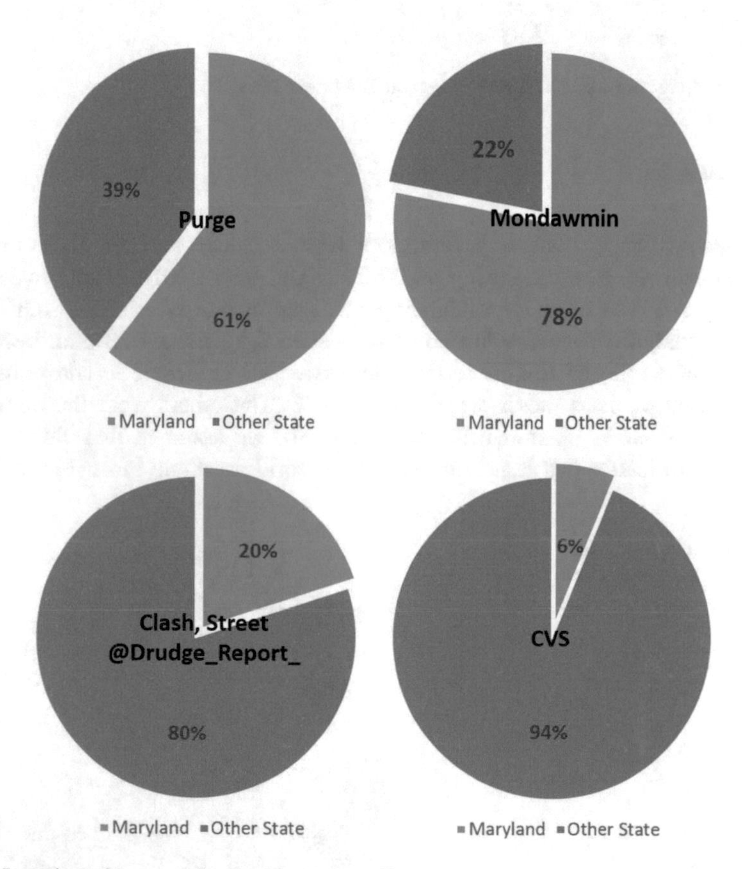

Fig. 6 Location of tweets related with mass media

For understanding the role of mass media in detail, we could get meaningful words based on TF-IDF at 5-min interval. As shown in the Fig. 6, we could choose words as important incidents such as purge, Mondawin, clash, streets, @Druge_Report_, and CVS because the result has very high scores compared with other words. Especially, during the period, students of high school gathered at Mondawim mall due to purge movie, and then riot was begun by students, and then CVS was fired.

First, during 2:25 p.m.–3:20 p.m., Purge was the most important word among words from TF-IDF at 2:26 p.m. At that time, Jessica Anderson, reporter of The Baltimore Sun, wrote a first tweet about purge. The tweet was 'Student 'purge' threat shuts down Baltimore businesses', and then Kevin Rector, crime reporter of the Baltimore Sun, at 2:28 p.m. and Mark Puente, investigative reporter of the Baltimore Sun, at 2:29 p.m. did a retweet from Jessica Anderson, and then Shimon Prokeuperz, CNN Producer, did a retweet at 2:56 p.m. Official Twitter of the Baltimore Sun wrote a tweet about purge at 2:30 p.m. At that time, the number of the total buzz was 170, and the number of tweets including purge was 120. However, the number of tweets including purge related with mass media was 107.

Based on user profiles of tweets, tweets related with mass media were 89 %, and tweets related with Maryland and Baltimore were 78 % as shown in the Fig. 6. As a result, local mass media, The Baltimore Sun, leaded a social issue about the initial purge incident, and Twitter users shared and exchanged local mass media's tweets as retweets, and tweets were exchanged in local area, Maryland and Baltimore, and it implies that it was one of the local social issues at that time based on analyzed tweets.

Second, during 3:20 p.m.–3:35 p.m., based on TF-IDF Mondawmin is one of the important words. Especially, a Twitter user wrote the first tweet of Mondawmin at 2:32 pm. However, The Baltimore Sun wrote a tweet of the first mass media related with Mondawmin at 3:30 p.m. The tweet was 'Heavy police presence at Mondawmin Mall, which has closed. Transit hub there is many students' way home from school', and then Malieka Flippen,announcer for radio One Baltimore, updated a tweet at 3:25 p.m. and then DanDuring, Baltimore Sun columnist, updated a tweet at 3:25 p.m. During this period, whole buzz was 133, and the number of tweets related with Mondawmin was 69, and tweets related with mass media was 66. As a result, in fact, although the first Mondawmin tweet was one of the Twitter users, the tweet didn't retweet to anyone, and local mass media, The Baltimore Sun, was a first writer related with mass media including Mondawmin, and we could not see any major mass media tweets. As a result, although a first writer about Mondawmin was one of the Twitter users, local mass media leaded to Mondawmin incident.

The percentage of tweets of local mass media related with Mondawmin was 96 %. Based on user profiles of tweets, tweets related with mass media were 96 %, and tweets related with Maryland and Baltimore were 61 % as shown in the Fig. 6. As a result, Twitter users shared and exchanged tweets about local mass media, and tweets shared and exchanged in local area, Maryland and Baltimore. Based on analyzed tweets, it implies that it was one of the local social issues.

Third, during 3:40 p.m.–4:00 p.m., based on TF-IDF, @Drudge_Report_, clash, and street were the most important words. As the result of user profiles, @Druge_Report is one of influential Twitter users. They have followers of 554 k. The Baltimore Sun has followers of 156 k. A tweet of @Druge_Report was 'New #clashes in #streets; #businesses shut down…'. During this period, the total number of buzz was 493, and the number of tweets of mass media including @Drudge_Report_ was 238, and tweets of @Drudge_Report_ were 176. Based on location information of tweets, the number of tweets written by other states were 80 %. As a result, from this period, because of @Drudge_Report_ tweets, tweets spread throughout the other states.

During 4:40 p.m.–7:30 p.m., CVS was one of the most important words using TF-IDF. Tweets written by other states related with CVS were 94 %, and tweets related with mass media were 1,123 among the total number of 25,208 tweets. Due to the CVS incident, people shared and exchanged their tweets to other users.

Based on user profiles of tweets, tweets written by other state were 94 %.

As a result, because of local mass media, it was one of the local social issues until the Mondawin incident, and then after the appearance of a influential twitter user in the online world, @Drudge_Report, it was one of major social issues at that time in United State.

5 Conclusion

This study aimed to address how the mass media intensively is related to the social media. Especially, we discover the relation between them based on the agenda setting theory.

As the result from our research, the mass media related Twitter users have a strong influence on social issues. It affects that the other users as audiences would be interested in the issues. We extract meaningful keywords using a method, TF-IDF. Then, we selected tweets generated in specific time when the keyword were exchanged. Then, we found out the role of mass media according to agenda setting theory into the social issue which is "Baltimore Riot". We found two characteristics to support agenda setting in a case study which is "Baltimore Riot". The first one is that mass media announced and leaded important incidents at initial period of the issue. Then, Twitter users exchanged and shared tweets related the issue. The second one is that the location of Twitter users located in local areas which are Baltimore and Maryland in the initial period. Then, the other users located in the other location exchanged and shared tweets about the issues. We assume that the influence users who have numerous followers or mass media related users affect the number of tweets. It implies that the agenda setting theory is existing in the social media.

References

1. Kwak, H., Lee, C., & Park, H. Sue Moon, What is Twitter, a Social Network or a News Media?
2. https://about.twitter.com/company.

3. McCombs, M., & Reynolds, A., (2002). News influence on our pictures of the world. Media effects: Advances in theory and research.
4. Lee, K., & Palsetia, D. Twitter Trending Topic Classification.
5. Kim, Y., Kim, S., Jaimes, A., & Oh, A. A Computational Analysis of Agenda Setting.
6. Camaj, L. Need for orientation, selective exposure, and attribute agenda-setting effects.
7. Wartena, C., Slakhorst, W., & Wibbels, M. Selecting keywords for content based recommendation.
8. Manning, C. D., Surdeanu, M., Bauer, J., Finkel, J., Bethard, S. J., & McClosky, D. (2014). The Stanford CoreNLP Natural Language Processing Toolkit (pp. 55–60). *Proceedings of the 52nd Annual Meeting of the Association for Computational Linguistics: System Demonstrations*.
9. Lott, B. Survey of Keyword Extraction Techniques. UNM Education, 2012.
10. Lee, S., & Kim, H. (2008). News keyword extraction for topic tracking. *Fourth International Conference on Networked Computing and Advanced Information Management* (pp. 554–559). NCM'08, IEEE.
11. Manning, C. D., Raghavan, P., & Schutze, H. (2008). Scoring, term weighting, and the vector space model. Introduction to Information Retrieval. pp. 100.
12. Liu, F., Liu, F., & Liu, Y. (2008). Automatic keyword extraction for the meeting corpus using supervised approach and bigram expansion. *Proceedings of IEEE SLT.*
13. Matsuo, Y., & Ishizuka, M. (2004). Keyword extraction from a single document using word co-occurrence statistical information. *International Journal on Artificial Intelligence, 13*(1), 157–169.
14. Byun, C., Lee, H., Kim, Y., & Kim, K. Automated Twitter Data Collecting Tool and Case Study with Rule-based Analysis.
15. Twitter Developers. https://dev.twitter.com.

On the Prevalence of Function Side Effects in General Purpose Open Source Software Systems

Saleh M. Alnaeli, Amanda Ali. Taha and Tyler Timm

Abstract A study that examines the prevalence and distribution of function side effects in general-purpose software systems is presented. The study is conducted on 19 open source systems comprising over 9.8 Million lines of code (MLOC). Each system is analyzed and the number of function side effects is determined. The results show that global variables modification and parameters by reference are the most prevalent side effect types. Thus, conducting accurate program analysis or many adaptive changes processes (e.g., automatic parallelization to improve their parallelizability to better utilize multi-core architectures) becomes very costly or impractical to conduct. Analysis of the historical data over a 7-year period for 10 systems shows that there is a relatively large percentage of affected functions over the lifetime of the systems although trend is flat in general, thus posing further problems for inter-procedural analysis.

Keywords Function side effects · Pass by reference · Function calls · Static analysis · Software evolution · Open source systems

1 Introduction

It is very challenging to statically analyze or optimize programs that have functions with side effects. Most studies show functions with side effects poses a greater challenge in many software engineering and evolution contexts including system

S.M. Alnaeli (✉) · A.Ali.Taha · T. Timm
Department of Computer Science, University of Wisconsin-Fox Valley,
Menasha, WI 54952, USA
e-mail: saleh.alnaeli@uwc.edu

A.Ali.Taha
e-mail: RICEA9147@studnest.uwc.edu

T. Timm
e-mail: TIMMT4191@studnest.uwc.edu

R. Lee (ed.), *Software Engineering Research, Management and Applications*, Studies in Computational Intelligence 654,
DOI 10.1007/978-3-319-33903-0_9

maintainability, comprehension, reverse engineering, source code validation, static analysis, and automatic transformation and parallelization.

For development teams, identifying functions that have side effects is critical knowledge when it comes to optimizing and refactoring software systems due to many reasons including regular adaptive maintenance or for more efficient software. One example is in the context of automatic parallelization (transformation of sequential code to a parallel one that can efficiently work on multicore architectures), where a for-loop that contains a function call with a side effect is considered un-parallelizable, e.g., cannot be parallelized using Openmp. It has been proved that the mosat prevalent inhibitor to parallelization is functions called within for-loops that have side effects. That is, this single inhibitor poses the greatest challenge in re-engineering systems to better utilize modern multi-core architectures.

A side effect can be produced by function call in multiple ways. Basically, any modification of the non-local environment is referred to as side effect [12, 18] (e.g., modification of a global variable or passing arguments by reference). Moreover, a function call in a for-loop or in a call from that function can introduce data dependence that might be hidden [15]. The static analysis of the body of the function increases compilation time; hence this is to be avoided. In automatic parallelization context, in spite the fact that parallelizing compilers, such as Intel's [13] and gcc [11], have the ability to analyze loops to determine if they can be safely executed in parallel on multi-core systems, they have many limitations. For example, compilers still cannot determine the thread-safety of a loop containing external function calls because it does not know whether the function call has side effects that would introduce dependences.

As such, it is usually left to the programmer to ensure that no function calls with side effects are used and the loop is parallelized by explicit markup using an API. There are many algorithms proposed for side-effect detection [5, 18], with varying efficiency and complexity.

In general, a function has a side effect due to one or more of the following:

1. Modifies a global variable
2. Modifies a static variable
3. Modifies a parameter passed by reference
4. Performs I/O
5. Calls another function that has side effects

When it comes to side effects detection, the problem gets worse if indirect calls via function pointers or virtual functions are involved. It is very challenging to statically analyze programs that make function calls using function pointers [6, 17] and virtual methods [4]. A single function pointer or virtual method can alias multiple different functions/methods (some of which may have side effects) and determining which one is actually called can only be done at run time. An imprecise, but still valid, analysis is to resolve all function pointers in the system and then assume that a call using each function pointer/virtual method reaches all possible candidate functions in the program. This, of course, adds more complexity and inaccuracy to static analysis. In general, the problem has been shown to be

NP-hard based on the ways function pointers are declared and manipulated in the system [6, 17, 19].

In spite of the fact that a variety of studies have been done using inter-procedural and static analysis on function side effects, to the best of our knowledge, no historical study has been conducted on the evolution of the open-source systems over time on regard to function side effect types distribution on open source systems. We believe that an extensive comprehension of the nature of side effect types distribution is needed for better understanding of the problem and its obstacles that must be considered when static analysis is conducted. We believe that development teams who may plan for conducting system refactorings for eliminating function side effects due to whatever reason (e.g., adapting the system for better exporting of multicore architectures) they should know what is the most prevalent type so that they consider starting with it for gaining better and optimal results and outputs from their work.

In this study, 19 large-scale C/C++ open source software systems from different domains are examined. For each system, the history of each system was examined based on multiple metrics. The number of functions with side effects are determined for each release. Then each kind of function side effect is determined (pass by reference, modification of global variable, in/out operation, and calling affected function). A count of all types are found and kept.

This data is presented to compare the different systems and uncover trends and general observations. The trend of increasing or decreasing numbers of side effect type is then presented.

We are particularly interested in determining the most prevalent function side effect type that occur in most of general open source applications, and if there are general trends. This work serves as a foundation for understanding the problem requirements in the context of a broad set of general purpose applications.

We are specifically interested in addressing the following questions. How many functions and methods in these systems do not have any side effects? Which types of side effects are most frequent? Understanding which side effect occur alone in functions is also relevant. That is very important for many software engineering context, e.g., automatic parallelization where functions side effects creates a well-known inhibitor to parallelization [1, 2]. Additionally, we propose and provide some simple techniques that can help avoiding and eliminating the function side effects, thus improving overall system maintainability, analyzability, and parallelizability.

This work contributes in several ways. First, it is one of the only large studies on the function side effects distribution and evolution on open source general purpose software systems. Our findings show that modification of global variables and parameters sent by reference represent the vast majority of function side effect types occurring in these systems. This fact will assist researchers in formulating and directing their work to address those problems for better software analysis, optimization, and maintenance in many recent software engineering contexts including source code transformation and parallelization.

The remainder of this paper is organized as follows. Section 2 presents related work on the topic of function side effects. Section 3 describes the functions with side effects and approached used for the detection and determination, along with all possible limitations in our approach. Section 4 describes the methodology we used in the study along with how we performed the analysis to identify each side effect. Section 5 presents the data collection processes. Section 6 presents the findings of our study of 19 open source general purpose systems, followed by conclusions in Sect. 6.

2 Related Work

There are multiple algorithms used for identifying and detecting function side effects. Our concern in this study is the distribution of side effect types and how software systems evolve overtime with respect to function side effect presence, in particular for general purpose large-scale open-source software systems, for better understanding and uncovering any trends or evolutionary patterns. That is, we believe that can be a valuable information in determining and predicting the solutions and effort required to better statically analyze those systems written in C/C++ languages and design efficient tools that can help eliminating those side effects for better quality source code.

The bulk of previous research on this topic has focused on the impact of side effects on system maintainability, analyzability, code validation, optimization, and parallelization. Additionally, Research continues to focus on improving the efficiency of interprocedural techniques and analyzing the complexity of interprocedural side effect analysis [16].

However, no study has been conducted on the evolution of the open source systems over time with respect to the presence of function side effects and their distribution on source code level as we are going to conduct in this work by examining the history of subset of systems for each release for 5-year period.

Cooper and Kennedy [10] conducted a study that shows a new method for solving Banning's alias-free flow-insensitive side-effect analysis problem. The algorithm employs a new data structure, called the binding multi-graph, along with depth-first search to achieve a running time that is linear in the size of the call multi-graph of the program. They proved that their method can be extended to produce fast algorithms for data-flow problems with more complex lattice structures. The study focused on the detection of side effects but did not provide any statistics about the usage and distribution of function side effects on the systems they studied and all software systems in general.

In source code parallelization context, most of compilers still cannot determine the thread-safety of a loop containing external function calls because it does not know whether the function call has side effects that would introduce dependences. That is, parallelizing compilers, such as Intel's [13] and gcc [11], have the ability to analyze loops to determine if they can be safely executed in parallel on multi-core systems, multi-processor computers, clusters, MPPs, and grids. The main limitation is effectively analyzing the loops when it comes to function side effects especially if function pointers or virtual functions are involved [16].

Alnaeli et al. [2] conducted an empirical study that examines the potential to parallelize large-scale general-purpose software systems. They found that the greatest inhibitor to automated parallelization of for-loops is the presence of function calls with side effects and they empirically proved that this is a common trend. They recommended that more attention needs to be placed on dealing with function-call inhibitors, caused by function side effects, if a large amount of parallelization is to occur in general purpose software systems so they can take better advantage of modern multicore hardware.

However, they have not provided and results that show the distribution of function side effect types in general purpose software systems. The work presented here differs from previous work on open source general purpose systems in that we conduct an empirical study of actual side effects in the source code level and all the potential challenges in this process. We empirically examine a reasonable number of systems, 19, to determine what is the most prevalent function side effect present in open source systems and how open source systems evolve over time with respect to function with side effects.

3 Functions Side Effects

We now describe the way we determine and detect side effects and the approach we followed in dealing with indirect calls that are conducted via function pointers and virtual methods within all detected functions. In general, any execution or interaction with outside world that may make the system run into unexpected status is considered a side effect. For example, any input/output operation conducted within a called function, or modification of the non-local environment is referred to as side effect [12, 18] (e.g., modification of a global variable or passing arguments by reference).

In this study, a function considered to have a side effect if it contains one or more of the following: (modification of a global variable, modification of a parameter passed by reference, I/O operation, or calling another function that has side effects.)

3.1 Determining Side Effects

To determine if a function/method has a side effect we do static analysis of the code within the function/method. We basically used the same approach in our previous studies, however in this study no user defined function/method is excluded [1, 2]. Any variables that are directly modified via an assignment statement (e.g., x = x + y) are detected by finding the l-value (left hand side variables) of an expression that contains an assignment operator, i.e., =, +=, etc. For each l-value variable it is determined if it has a local, non-static declaration, or is a parameter that is passed by value. If there are any l-value variables that do not pass this test, then the function is labeled as having a side effect. That is, the function is modifying either a global, static, or reference parameter. This type of side effect can be determined with 100 % accuracy since the analysis is done local to the function only.

Of course, pointer aliasing can make detecting side effects on reference parameters and global variables quite complex. Our approach detects all direct pointer aliases to reference parameters and globals such as a pointer being assigned to a variable's address (int *ptr; ptr = &x;). If any alias is an l-value we consider this to cause a side effect. However, we currently do not support full type resolution and will miss some pointer variables. Also, there are many complicated pointer aliasing situations that are extremely difficult to address [14] even with very time consuming analysis approaches. For example, the flow-sensitive and context-sensitive analysis algorithms can produce precise results but their complexity, at least O(n3), makes them impractical for large systems [14]. As such, our approach to detection of side effects is not completely accurate in the presences of pointer aliasing. However, this type of limited static pointer analysis has shown [3] to produce very good results on large open source systems.

The detection of I/O operations is accomplished by identifying any calls to standard library functions (e.g., printf, fopen). A list of known I/O calls from the standard libraries of C and C++ was created. Our tool checks for any occurrence of these and if a function contains one it is labeled as having a side effect. Also, standard (library) functions can be labeled as side effect free or not. As such, a list of safe and unsafe functions is kept and our tool checks against this list to further identify side effects.

Our detection approach identifies all function/method calls within a caller function/method. The functions directly called are located and statically analyzed for possible side effects through the chain of calls. This is done for any functions in the call graph originating from the calls in the caller function/method. This call graph is then used to propagate any side effect detected among all callers of the function.

Even with our analysis there could still be some functions that appear to have side effects when none actually exists or that the side effect would not be a problem to parallelization. These cases typically require knowledge of the context and problem being addressed and may require human judgment (i.e., may not be automatically determinable). However, our approach does not miss detecting any potential side effects. As such we may over count side effects but not under count them.

3.2 Dealing with Function Pointers and Virtual Methods

Our approach for calls using function pointers and virtual methods is to assume that all carry side effects. It is the same approach we used in past studied but this time with more involved functions and methods regardless their locations in the system (my papers). At the onset, this may appear to be a problematic, however conservative, limitation. It is very challenging to statically analyze programs that make function calls using function pointers [6, 17] and virtual methods [4]. A single function pointer or virtual method can alias multiple different functions/methods and determining which one is actually called can only be done at run time. An imprecise, but still valid, analysis is to resolve all function pointers in the system and then assume that a call using each function pointer/virtual method reaches all possible candidate functions in the program. This, of course, adds more complexity and inaccuracy to static analysis. In general, the problem has been shown to be NP-hard based on the ways function pointers are declared and manipulated in the system [6, 17, 19].

Function pointers can come in various forms: global and local function pointers. Global forms are further categorized into defined, class members, array of function pointers, and formal parameters [17]. Our tool, *SideEffectDetector*, detects all of these types of function pointers whenever they are present in a function/method. Pointers to member functions declared inside C++ classes are detected as well. Classes that contain at least one function pointer and instances derived from them are detected. Locally declared function pointers (as long as they are not class members, in structures, formal parameters, or an array of function pointers) that are defined in blocks or within function bodies are considered as simple or typically resolved pointers.

Detecting calls to virtual methods is a fairly simple lookup. We identify all virtual methods in a class and any subsequent overrides in derived classes. We do not perform analysis on virtual methods, instead it is assumed that any call to a virtual has a side effect. Again, this is a conservative assumption and we will label some methods that in actuality do not have a side effect to be a problem. A slightly more accurate approach would be to analyze all variations of a virtual method and if none have side effects then it would be a safe call. However, this would require quite a lot of extra analysis with little overall improvement in accuracy.

4 Methodology for Detecting Function Side Effects

A function or method is considered a pure if it does not contain any side effect. We used a tool, ParaSta, developed by one of the main authors and used in [1]; Alnaeli et al. [2], to analyze functions and determine if they contain any side effect as defined in previous section. First, we collected all files with C/C++ source-code extensions (i.e., c, cc, cpp, cxx, h, and hpp). Then we used the srcML (www.srcML.org) toolkit (Collard et al. [7–9] to parse and analyze each file. The srcML format wraps the statements and structures of the source-code syntax with XML elements, allowing tools, such as SideEffectDetector, to use XML APIs to locate such things as function/method implementation and to analyze expressions. Once in the srcML format, SideEffectDetector iteratively found each function/method and then analyzed the expressions in the function/method to find the different side effects. A count of each side effect per function was recorded. It also recorded the number of pure functions found. The final output is a report of the number of pure (clear) functions and functions with one or more types of side effects. All functions were deeply analyzed and side effect types distributions were determined as well.

Findings are discussed later in this paper along with limitations of our approach.

5 Data Collection

Software tools were used, which automatically analyze functions and determines if they contain any side effects. The srcML toolkit produces an XML representation of the parse tree for the C/C++ systems we examined. SideEffectDetector, which was developed in C#, analyze the srcML produced using XML tools to search the parse tree information using system.xml from the.NET framework. The body of each function is then extracted and examined for each type of side effects in functions. For the function, if no side effect exists in a function it is counted as a pure function otherwise the existence of each side effect is recorded. The systems that were chosen in this study were carefully selected to represent a variety of general purpose open source systems developed in C/C++. These are well-known large scale systems to research communities.

6 Findings, Results, and Discussion

We now study the distribution of 19 general purpose open-source software projects. Table 1 presents the list of systems examined along with number of files, functions, and LOCs for each of them.

Table 1 The 19 open source systems used in the study

System	Language	KLOC	Functions	Files
gcc.3.3.2	C/C++	1,300,000	35,566	10,274
TAO	C++	1,543,805	39,720	10,000
openDDS3.8	C++	326,471	9,185	1,779
ofono	C	242,153	7,331	527
GEOS	C++	173,742	5,738	815
CIAO	C++	191,535	5,937	1,044
DanCE	C++	102,568	5,292	345
xmlBlaster	C/C++	92,929	2,590	429
Cryto++	C++	70,365	3,183	274
GMT	C++	261,121	4,204	290
GWY	C++	392,130	8,340	594
ICU	C++	825,709	15,771	1,719
KOFFICE	C/C++	1,185,000	40,195	5,884
LLVM	C/C++	736,000	27,922	1,796
QUANTLIB	C++	449,000	12,338	3,398
PYTHON	C	695,000	12,824	767
OSG	C++	503,000	15,255	1,994
IT++	C++	120,236	4,220	394
SAGAGIS	C++	616,102	11,422	1,853
TOTAL	–	9,826,102	89,180	18,215

These systems were chosen because they represent a variety of applications including compilers, desktop applications, libraries, a web server, and a version control system. They represent a set of general-purpose open-source applications that are widely used. We have a strong feeling gained from their popularity in literature that they represent a good reflection of the types of systems that would undergo reengineering or migration to better take advantage of available technologies and architectures, and targeted for regular maintenance, parallelization, evolution, and software engineering processes in general.

6.1 Design of the Study

This study focuses on three aspects regarding side effect type distribution and evolution in general purpose systems. First, the percentage of functions containing one or more side effects. Second, we examine which side effect type are most prevalent. Third, we seek to understand the cause of parallelizability inhibitors and as a case study we focus on function side effects.

Third, we seek to understand the when side effects are the sole cause in function affection. That is, function can have multiple side effects and therefore would require a large amount of effort to remove all the side effects. Thus we are interested in understanding how often only one type of side effect occurs in a functions. These types of functions would hopefully be easier to refactor into something that is pure and simple to analyze. Lastly, we examine how the presence of side effects changes over the lifetime of a software system. We propose the following research questions as a more formal definition of the study:

RQ1: What is a typical percentage of function that are pure and clear from any side effects (have no side effects)?
RQ2: Which types of side effects are the most prevalent?
RQ3: What their distributions? Exclusively and inclusively (affected by only one type of side effects)
RQ4: Over the history of a system, is the presence of function side effects increasing or decreasing?

We now examine our findings within the context of these research questions.

6.2 Percentage of Pure Functions (Has No Side Effects)

Figure 1 presents the results collected for the 19 open source systems. We give the total number of functions that have side effects along with the percentage of all side effects we detected. Figure 1 shows the percentage of affected functions computed over the total number of function. As can be seen from 1, affected functions account

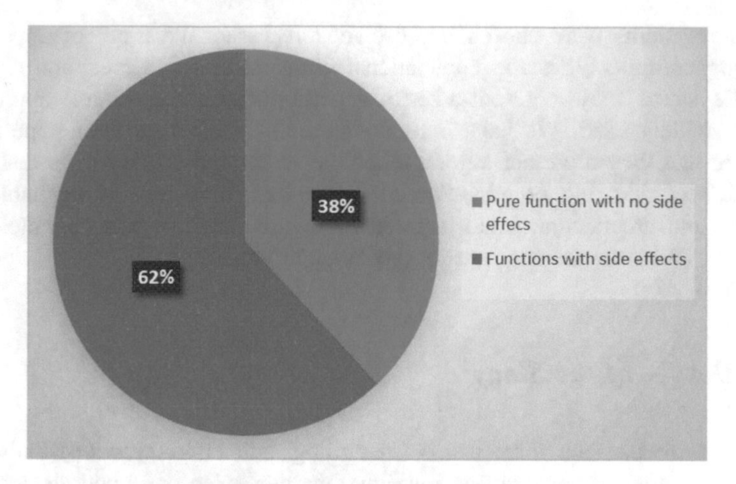

Fig. 1 total average of pure functions versus affected functions in all 19 systems

for between 29 and 95 % of all function and methods in these systems, with an overall average of 62 %. However, in general, the percentage is high for most of the systems. That is, on average, a big portion of the detected functions in these open source systems could hold side effects. This addresses RQ1.

6.3 Distribution of Function Side Effects

We now use our finding to address RQ2. 1 present the details of our findings on the distribution of side effects in studied open source systems. It clearly the counts of each side effect that occur within functions. Many of the functions have multiple side effects (e.g., a pass by reference and modification of global variables). As can be seen, modification of global variables is by far the most prevalent across all systems.

For most of the systems this is then followed by passing parameters by reference and then calling a function that has a side effect, thus addressing R2. The findings show that DanCE has the lowest percentage which make it a model system when it comes to function side effects addressing through development processes. In contrast, Python seems to have big challenges when it comes to function side effects.

Figures 1 and 2 give the percentage of functions that contain only one type of side effects for each category (addressing R3). The average percentage is also given and this indicates that functions that have modification of global variable are clearly more prevalent. We see that CIAO has the largest percentage of modification of global variables as a side effect 25 %, followed by IT++ at 22 %. GWY has the lowest, at 3 %. The percentage of the for-loops that contain only a I/O operation across all the systems is quite small by comparison. Interesting fact here is that passing parameters by reference is considerably high as well for multiple systems as shown in Fig. 3 and Table 2.

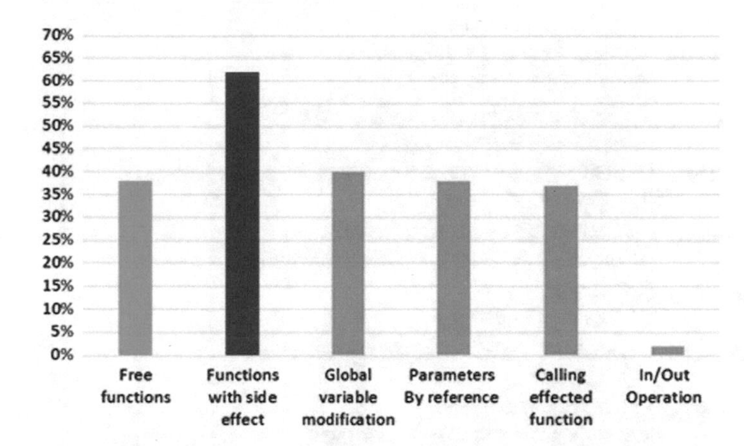

Fig. 2 Average Percentage of functions in all systems that contain only a single type of side effects. The remaining functions are either pure or have multiple types of side effects

6.4 Historical Change of Function Side Effect Frequency

Now we present a historical study we conducted on a subset of ten systems chosen from the 19 studied systems, for 7-year period. Each of those systems, has been under development for 7 years or more. To address R4, we examined the versions from 2005 to 2011 (7-year period those 10 systems). Our goal is to uncover how each system evolves in the context of function purity and complexity. Here we measure this by examining the change of side effects within function/methods. Our feeling is that this information could lead to recommendations for utilizing and adapting to the current software and hardware trends.

The change in the percentage of affected function with side effects, and presences of each side effect was computed for each version in the same manner as we described in the previous sections.

These values were aggregated for each year so the systems could be compared on a yearly basis. The systems were updated to the last revision for each year. As before, all files with source code extensions (i.e., c, cc, cpp, cxx, h, and hpp) were examined and their functions were then extracted. Figure 4 presents the change in the percentage of affected functions for each of the 10 systems.

During the 7-year period all systems show a fairly flat trend during the duration. Two systems, OSG and Crypto++, have a steep decline at the end of the period. Geos, Loki, and Koffice have increase for about one year early on and then are relatively flat in proceeding years. Figure 5 presents the percentage of functions that contain a modification of global variable as a side effect. It is approximately a same trend of Fig. 6.

Figure 7 presents the change in the percentage of affected functions by passing arguments by reference for each of the 10 systems. During the 7-year period all systems show a fairly flat trend during the duration except for GMT which starts to

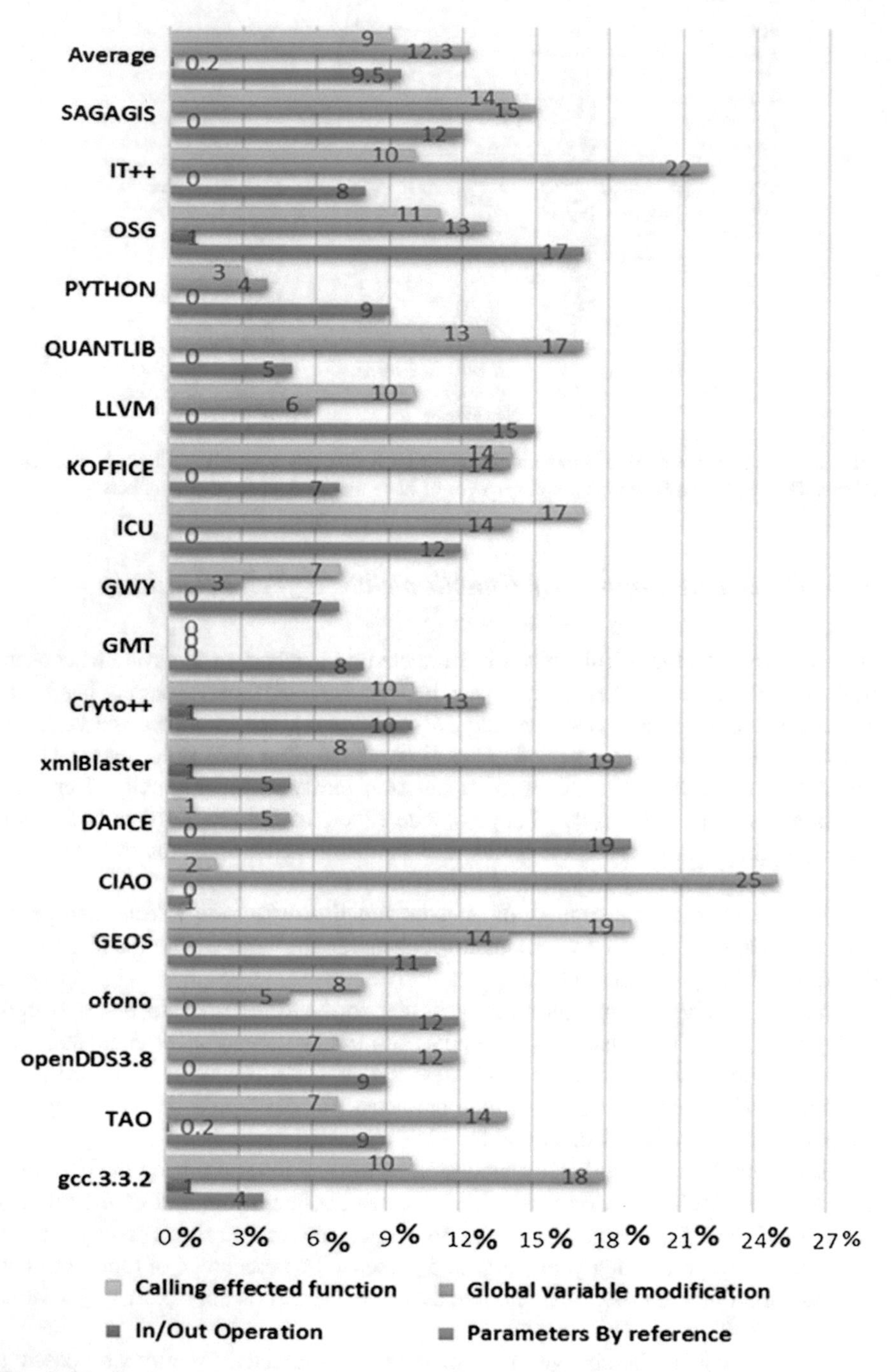

Fig. 3 Distribution of Percentage of functions in all systems that contain only a single type of side effects. The remaining functions are either pure or have multiple types of side effects

Table 2 Side effects distribution the 19 open source systems used in the study

System	Functions with side effects	Parameters by reference	In/out operation	Global variable modification	Calling effected function
gcc.3.3.2	18,803 (52 %)	7,098 (19 %)	1,728 (4 %)	12,685 (35 %)	11,015 (30 %)
TAO	20,349 (51 %)	11,451 (28)	361 (<1 %)	13,315 (33 %)	8,135 (20 %)
openDDS3.8	4,414 (48 %)	2,398 (26 %)	244 (2 %)	2,815 (30 %)	2,026 (22 %)
ofono	6,822 (93 %)	5,816 (79 %)	61 (<10 %)	5,283 (72 %)	5,256 (71 %)
GEOS	3,638 (63 %)	1,684 (29 %)	63 (1 %)	1,840 (32 %)	2,515 (43 %)
CIAO	2,187 (36 %)	526 (8 %)	18 (<1 %)	1,953 (32 %)	405 (6 %)
DanCE	1,584 (29 %)	1,217 (22 %)	4 (<1 %)	459 (8 %)	165 (3 %)
xmlBlaster	1,415 (54 %)	630 (24 %)	131 (5 %)	988 (38 %)	810 (31 %)
Cryto++	1,700 (53 %)	834 (26 %)	96	950 (29 %)	1,084 (34 %)
GMT	3,961 (94 %)	3,886 (92 %)	86 (2 %)	3,559 (86 %)	1,919 (59 %)
GWY	7,524 (90 %)	6,619 (79 %)	12 (<1 %)	6,261 (75 %)	5,056 (30 %)
ICU	11,545 (73 %)	6,388 (40 %)	302 (1 %)	6,801 (43 %)	8,572 (54 %)
KOFFICE	21,259 (52 %)	9,109 (22 %)	205 (<1 %)	11,918 (29 %)	13,682 (34 %)
LLVM	15,135 (54 %)	10,080 (36 %)	76 (<1 %)	7,804 (27 %)	10,289 (36 %)
QUANTLIB	6,242 (50 %)	2,344 (18 %)	96 (<1 %)	3,772 (30 %)	3,427 (27 %)
PYTHON	12,282 (95 %)	11,081 (86 %)	183 (1 %)	10,466 (81 %)	10,156 (79 %)
OSG	10,160 (66 %)	6,035 (39 %)	376 (2 %)	5,464 (35 %)	6,643 (43 %)
IT++	2,562 (60 %)	1010 (23 %)	174 (4 %)	1,712 (40 %)	1,646 (39 %)
SAGAGIS	7,089 (62 %)	3,634 (31 %)	65 (<1 %)	3,948 (34 %)	4,715 (41 %)
Average	62 %	38 %	2 %	40 %	37 %

increase by mid of 2010 to reach about 90 %. Loki has a steep decline early on and then are relatively flat in proceeding years.

Figure 3 presents the percentage of functions that contain a side effect caused by calling another function that has side effect. Figure 3 shows that a flat trend for most of the systems except for Loki which has an identical trend with other side effects. That is, it is approximately a same trend of Figs. 4, 5, 6 and 7.

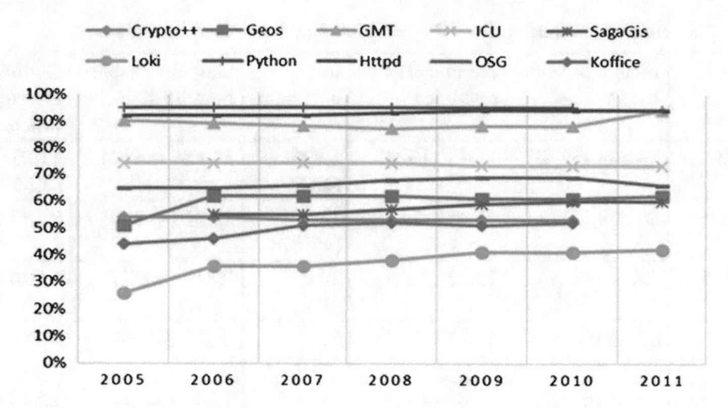

Fig. 4 The evolution of the percentage of functions that has side effects over a seven-year period for the ten systems

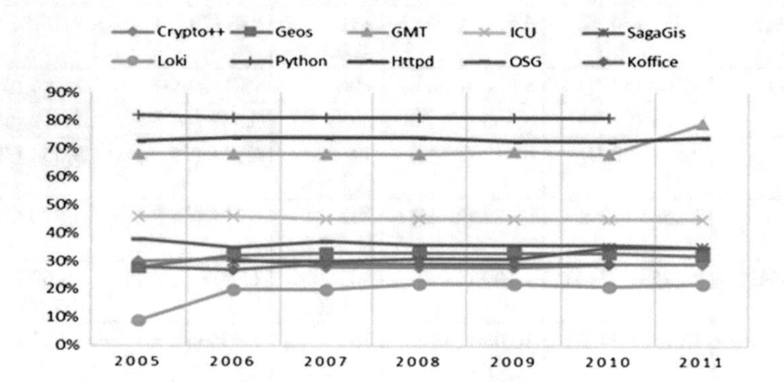

Fig. 5 The percentage of function that modify global variable over a seven-year period for the ten systems

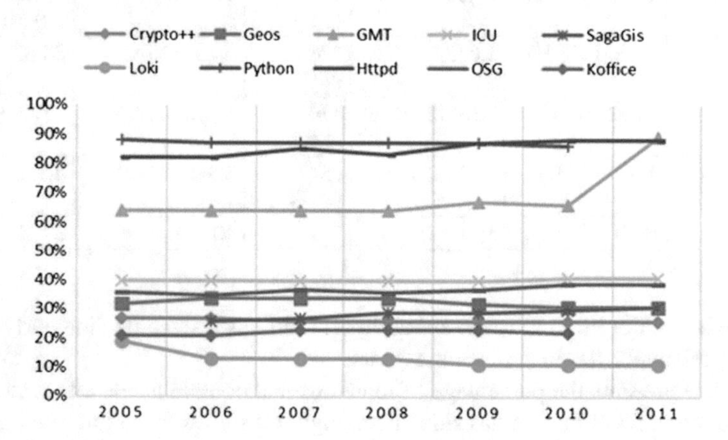

Fig. 6 The percentage of function that had parameters by reference over a seven-year period for the ten systems

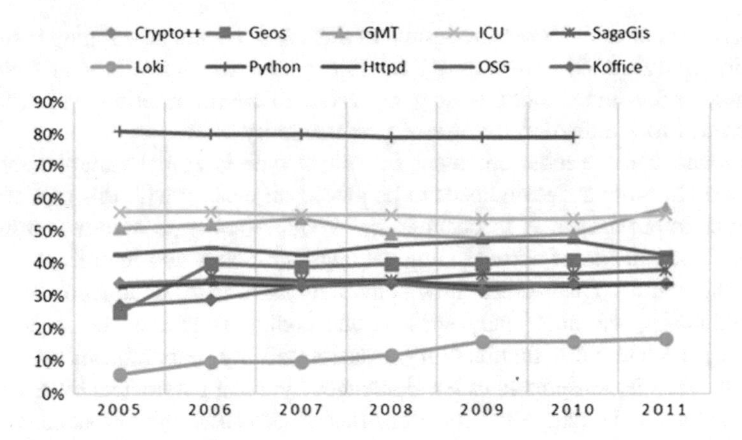

Fig. 7 The percentage of function that calls another affected function over a seven-year period for the ten systems

7 Conclusion

This study empirically examined the distribution and purity of functions (thus most of affected software engineering contexts by side effects) of 19 open source general purpose software systems from different system domains. There are no other recent studies of this type currently in the literature targeting function side effects in particular. We found that the greatest side effect of functions is the presence of modification to global variables followed closely by passing parameters by reference. As such, more attention needs to be placed on dealing with those types of side effects if a large amount of flexibility and easiness is to occur in general purpose software systems so they can be adapted to many recent techniques (e.g., source code parallelization and optimization). While we cannot completely generalize this finding to all software systems (across all domains) there is some indication that this is a common trend.

From our findings we believe that most development teams and organizations have not focused on developing software in a way that has minimum use of side effects so that they could one day take advantage of many new technologies e.g., parallel architectures. In the parallelization context for example, the recent ubiquity of multicore processors gives rise to the need to educate developers and make them more aware of the problems that can greatly affect their source code. As we have shown in many studies (Alnaeli et al. [2]), coding style can play a big role in advancing a system's parallelizability. The software engineering community needs to develop standards and idioms that help developers to avoid the side effects.

The objective of this study was to better understand what obstacles are in place for advancing the reengineering of systems to better take advantage of software engineering techniques e.g., static analysis. We are particularly interested in tools that assist developers in an automated or semi-automated manner to refactor or transform functions that have side effects to pure versions that can facilitate static analysis by other tools. From the results of this work we are developing methods to assist in removing side effects e.g., sending parameters by values instead of by reference, and using structures sent by value to return multiple values from a function call to elimination the use of parameters by reference.

We found that the most prevalent side effect type is global variables modification. As such, more attention needs to be placed on dealing with this type if a large amount of improvement is to occur in open source software systems so they can take better advantage of software engineering techniques and the recent technologies in the market. Our results show some indication that this is a common trend.

Additionally, we empirically showed that coding style can play a big role in advancing a system's maintainability, transformability, parallelizability, and analyzability. That is, developers cause challenges by using parameters by reference in their functions and having their functions modify global variables that can be easily handled outside of the functions. That is at least to some extent due to development teams and organizations not focusing on developing software in a way that could be easily analyzed, comprehended, and maintained. However, the recent challenges in software systems in general when it comes to adaptive changes gives rise to the need to educate software developers and engineers and make them aware of the problems that may be caused by using side effects when developing functions.

Acknowledgments This work was supported in part by a grant from the professional development program at University of Wisconsin-Fox Valley and UW-Colleges.

References

1. Alnaeli, S. M., Alali, A., & Maletic, J. I. (2012). Empirically Examining the Parallelizability of Open Source Software System. In *Proceedings of the 2012 19th Working Conference on Reverse Engineering* (pp. 377–386). IEEE Computer Society.
2. Alnaeli, S., Maletic, J., & Collard, M. (2015). An empirical examination of the prevalence of inhibitors to the parallelizability of open source software systems. *Empirical Software Engineering*, 1–30.
3. Alomari, H. W., Collard, M. L., Maletic, J. I., Alhindawi, N., & Meqdadi, O. (2014). srcSlice: very efficient and scalable forward static slicing. *Journal of Software: Evolution and Process*.
4. Bacon, D. F., & Sweeney, P. F. (1996). Fast static analysis of C++ virtual function calls. *SIGPLAN Not., 31*(10), 324–341.
5. Banning, J. P. (1979). An efficient way to find the side effects of procedure calls and the aliases of variables. In *Proceedings of the 6th ACM SIGACT-SIGPLAN Symposium on Principles of Programming Languages* (pp. 29–41). San Antonio, Texas: ACM.
6. Cheng, B.-C., & Hwu, W. (2000). An empirical study of function pointers using SPEC benchmarks. In *Proceedings of the 12th International Workshop on Languages and Compilers for Parallel Computing* (pp. 490–493). Springer-Verlag.
7. Collard, M. L., Maletic, J. I., & Marcus, A. (2002). Supporting document and data views of source code. In *Proceedings of ACM Symposium on Document Engineering* (p. 8).
8. Collard, M. L., Kagdi, H. H., & Maletic, J. I. (2003). An XML-based lightweight C++ fact extractor. In *11th IEEE International Workshop on Program Comprehension, 2003*.
9. Collard, M. L., Decker, M. J., & Maletic, J. I. (2011). Lightweight Transformation and Fact Extraction with the srcML Toolkit. In *Proceedings of the 2011 IEEE 11th International*

Working Conference on Source Code Analysis and Manipulation (pp. 173–184). IEEE Computer Society.
10. Cooper, K. D., & Kennedy, K. (1988). Interprocedural side-effect analysis in linear time. *SIGPLAN Not., 23*(7), 57–66.
11. Feng, L. (2009). *Automatic parallelization in graphite*. Retrieved from http://gcc.gnu.org/wiki/Graphite/Parallelization.
12. Ghezzi, C., & Jazayeri, M. (1982). *Programming language concepts.* Wiley.
13. Intel. (2010). *Automatic parallelization with intel compilers*. Retrieved from http://software.intel.com/en-us/articles/automatic-parallelization-with-intel-compilers/.
14. Mock, M., Atkinson, D. C., Chambers, C., & Eggers, S. J. (2005). Program slicing with dynamic points-to sets. *IEEE Transactions on Software Engineering, 31*(8), 657–678.
15. Oracle. (2010). *Subprogram call in a loop*. Retrieved from http://docs.oracle.com/cd/E19205-01/819-5262/aeuje/index.html.
16. Richardson, S., & Ganapathi, M. (1987). *Interprocedural analysis useless for code optimization*. Stanford University.
17. Shah Anand, R. B. G. (1995). Function pointers in C—An empirical study. Technical report LCSR-TR-244, p. 11.
18. Spuler, D. A., & Sajeev, A. S. M. (1994). Compiler detection of function call side effects. Technical report 94/01.
19. Zhang, S., & Ryder, B. G. (1994). *Complexity of single level function pointer aliasing analysis*. Rutgers University, Department of Computer Science, Laboratory for Computer Science Research.

Object Oriented Method to Implement the Hierarchical and Concurrent States in UML State Chart Diagrams

E.V. Sunitha and Philip Samuel

Abstract The event driven systems can be modeled and implemented using UML state chart diagrams. Code generation tools are used in the software development for making software system designs and for automatically generating skeletal source code from the system designs. Many research works concentrate on the automatic code generation from the state diagrams. Unfortunately the existing Object oriented languages do not support the direct implementation of state diagrams. We cannot find a one to one mapping between elements in the state chart diagram and the Object oriented programming constructs. The two main components of state diagram that cannot be effectively implemented in object oriented way is state hierarchy and concurrency. In this paper, we present an implementation pattern for the state diagram which includes both hierarchical and concurrent states. The state transitions of parallel states are delegated to the composite state class. We implemented the proposed approach and compared with similar tools and the result is promising.

Keywords Code generation · State machine · MDD · Executable UML

1 Introduction

Object Orientation plays an important role in software development. It represents a problem domain as a set of interacting objects coming under the problem domain. Object orientation made a revolution in analysis, design, and implementation

---

E.V. Sunitha (✉)
Department of Computer Science, Cochin University of Science & Technology, Kochi, India
e-mail: sunithaev@gmail.com

P. Samuel
Division of IT, School of Engineering, Cochin University of Science & Technology, Kochi, India
e-mail: Philips@cusat.ac.in

R. Lee (ed.), *Software Engineering Research, Management and Applications*, Studies in Computational Intelligence 654,
DOI 10.1007/978-3-319-33903-0_10

phases of the software development. Object Oriented Analysis (OOA) is used for developing an object model of the application domain. Object Oriented Design (OOD) develops an object oriented system model which satisfies the user requirements. Object Oriented Programming (OOP) implements the OOD using an OO programming language. This object orientation in all major phases of software development introduces increased understandability, maintainability, and reusability of design as well as code [1]. Hence, the object oriented methodology is advised for cost-effective, faster, and flexible software development.

UML is one of the designing languages which support object orientation in the design phase. It supports the important concepts of OOD such as, abstraction, inheritance, modularity, polymorphism etc. UML help us to design the structure as well as behavior of the system. They are called structural modeling and behavioral modeling. Structure diagram includes class diagram, object diagram, deployment diagram etc. Behavioral diagrams include activity diagram, state chart diagram, interaction diagram etc.

Similarly, OO programming languages like Java, C++, C# etc. are useful in the implementation phase. This helps us to continue the object orientation in the design phase to the implementation phase. There are some elements in UML design which can be directly mapped to any object oriented programming construct. Some elements in UML cannot be directly mapped to any programming element. Earlier, the designers design the system models using UML or other tools and given to the software engineers for coding. The software engineers had to start from the scratch, beginning from the inclusion of header files, declaration of variables etc. Over time, this scenario had been changed and there came some CASE tools, IDEs etc. for supporting the software engineers. These tools generate skeletal code from the designs we have modeled in UML or similar languages so that the programmer need not start from the scratch.

In the next generation of software development, there comes the Model Driven Development (MDD). MDD describes methods to develop software purely based on the system design. This concept leads us to a method of directly executing the system models even without converting them to the implementation code. It is called executable UML [2].

Many tools are available to convert the system designs to the source code. The system design may include class diagrams, state charts, activity diagrams, sequence diagrams etc. UML diagrams like class diagram can be easily converted to the source code because OO languages support the class concept. The class declaration statements, class definition statements, method definition statements, object creation, method invocation statement etc. are available in the existing OO languages. Some other diagrams like state chart diagram, activity diagram etc. cannot be directly mapped to OO program [3]. This is because of the lack of programming elements that can represent the elements in these diagrams.

In this paper we focus on state machines. State machines are useful to model event driven systems. It can represent the full life cycle of an object. It shows different states of the object and the transition between the states. We need to find out an efficient way to convert state charts to a program since there is no programming construct exist to directly represent elements in the state diagram.

In this paper we present a method to convert hierarchical states and concurrent states to Java code. In our method we follow a design pattern based approach. A design pattern gives the overall implementation outline using a class diagram. The surveys on code generation from state machines [4] show that the research outcomes are not giving an effective method to implement the concurrent states. We present a design pattern to implement both the state hierarchy and concurrency.

The main contributions of the paper are as follows:

- It presents a less complex design pattern for state machine implementation.
- It gives an effective method to implement Composite states with parallel regions.

The paper is organized as follows. Section 2 surveys different methods to convert state charts to programs. Section 3 introduces a new approach to convert the UML state chart diagrams to object oriented code. Section 4 presents a case study to demonstrate the proposed approach. Section 5 presents the related works. Section 6 describes the implementation and evaluation of the proposed approach and Sect. 7 concludes the paper.

2 Mapping State Charts to Programming—Different Approaches

In this section we discuss standard state machine implementation techniques which we can find in the literature. The methods include nested switch statement, state table, state design patterns, and UML meta class. The section ends with a proposal of new approach which supports hierarchical state machines and concurrent states. Figure 1 shows the basic elements in the UML state diagram.

2.1 Using Switch Statement

It is one of the simple and straight forward way of implementing simple state charts [5]. Here the state of the object is represented using a scalar variable and the events are represented as methods. Transitions are shown as the change in the state variables. Switch statements will be used for selecting the state and inside each case statement the internal, exit, and entry actions will be specified. There will be a context class corresponds to each state diagram and event methods will be the member functions of the same class.

Nested switch statements can be used instead of simple switch statement so that there is no need to represent different event methods. All the events can be handled

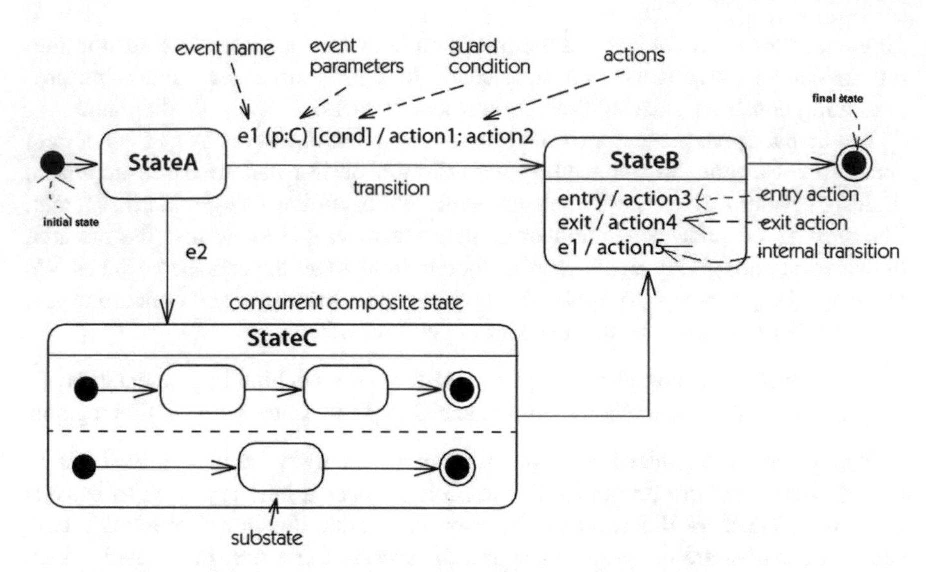

Fig. 1 State chart essentials (*Source* UML Reference manual, p. 527)

in one event method. The outer switch statement is used for state/event selection and the inner one is used for event/state selection. The transition logic will be given inside the case statement.

This method is suitable for simple state charts. As the state chart becomes complex, the code generated will be bulky. It will be difficult to modify the state chart because of the redundancy of code. Only OR-states can be represented in this method. AND-states cannot be included. This switch case approach does not support code reusability.

Jakimi [6] presents a different approach to implement state chart diagrams. Here, each state chart will be converted to a single class in the Java code. Different states in the state machine will be defined as the attributes of the implemented class, and the events will be represented as methods in the implemented class. For example, see the Fig. 2 where a simple state machine of an engine is shown. It has two states, idle and running. It is mentioned in the state diagram that the attribute on should be zero during idle state and on should be 1 during running state. In the implementation phase a single class, class Engine, is generated. The attributes that decides the state of the object becomes attributes in the generated class. Here it is the attribute on. The event that triggers the transition becomes a member function of the class. Here it is switchOn(). Inside the switchOn() function the value of the attributes will be updated according to the state machine.

2.2 Using State Design Patterns

In state design pattern approach, there will be a class diagram pattern that has to be followed for implementing all state chart diagrams [7]. There will be one class in the pattern which represents the context (domain) of the state chart diagram. The states in the state chart diagram are abstracted in a single abstract class which will act as an interface to the states in the state chart. The events will be the virtual member functions of the abstract state class. Each individual state in the state chart will be represented as the object of the derived class of abstract state class. If there are 'm' states in the state chart, then there will be 'm' different concrete state classes derived from the abstract state class. A sample state design pattern is shown in Fig. 2.

The object of the context class represents the domain object that needs to be represented in the program. The context class will have a data member (state variable) which represents the current state of the domain object. All the events are represented as member functions of the context class which in turn delegates the function to the corresponding state class objects.

Using state design patterns we can bring the object orientation in the state machine implementation. The domain object, whose state chart is drawn, is implemented as the object of the context class, each state of the domain object is implemented as the object of the corresponding concrete state class. Events are represented as the handles of the abstract state class and the transitions are accomplished by updating the state variables. This approach supports code reusability and avoids redundancy in coding.

There can be variable type of patterns that can be used to represent the state chart diagram. In the patterns, there is an abstract class which acts as an interface for the state classes. The interface will be connected to the context class. The pattern has an additional object called collaboration object to accomplish the sub states. It is an abstract class which acts as an interface for the sub states [8]. This object oriented approach creates some inconvenience too. In order to add a new state, we have to

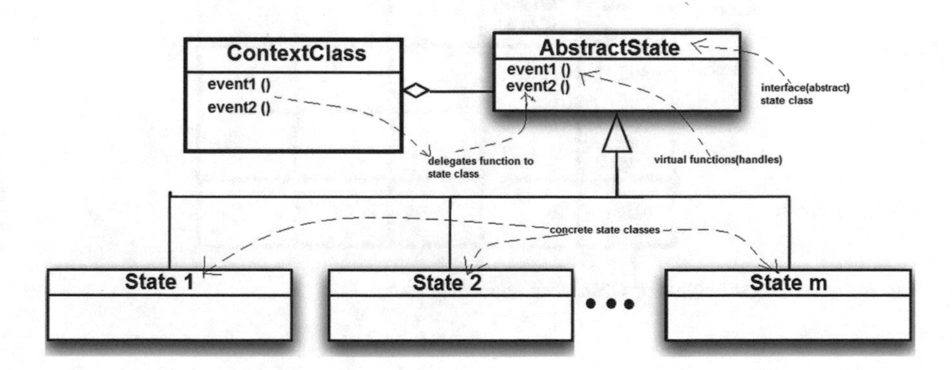

Fig. 2 Sample state design pattern

derive one more concrete state class from the abstract state class. Similarly, to add a new event, we need to add one more virtual function to the abstract state class.

2.3 *Using UML Meta Class*

Similar to state design pattern approach, the state chart can be mapped to the meta class structure defined for UML state machine [9]. That means, the implementation of the state chart uses a collection of related class. In this approach, state machine is represented as an object which is composed of 3 objects; State, Transition, and Events. Transition is again interconnected to State and Events.

Whenever a state machine is implemented it must include these four classes; State machine class, State class, Transition class, and Event class. Each state in the state diagram will be an object of the State class, likewise each event will be an object of the Event class and so.

2.4 *Using State Tables*

In this approach, the events and the states are entered in a table. The rows show the states and the columns show the events. The internal data of the tables shows the new state of the object when event 't' happens during the state 's' and the internal actions. Figure 3 gives the structure of a state table. The transitions are given inside the table. For example, if Event 1 occurs, when the object is in State 1, then action1 () will take place and the object will be changed to State 3.

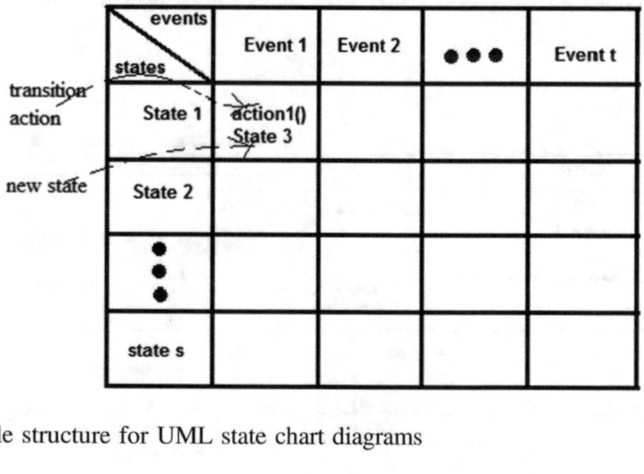

events / states	Event 1	Event 2	•••	Event t
State 1	action1() State 3			
State 2				
• • •				
state s				

transition action

new state

Fig. 3 State table structure for UML state chart diagrams

The state table can be directly used for coding. It makes the processing easier and faster. One drawback of this approach is that, it does not support object orientation. Moreover, the size of the table depends on the number of states and the number of events present in the state chart diagram. The table size does not depend on the number of transitions. Hence the table can be large even though the numbers of transitions are less. This in turn results in wastage of memory.

3 New Approach for Mapping State Chart to OO Program

The code generation from hierarchical and concurrent states in UML state machines are addressed in some research works [8, 10, 11]. Those methods give a complex design pattern to implement the state machine. In this paper, we propose a pattern based approach to convert the state machine to object oriented program. Figure 4 shows the proposed design pattern. Table 1 gives the mapping between state chart and the Object Oriented programming constructs.

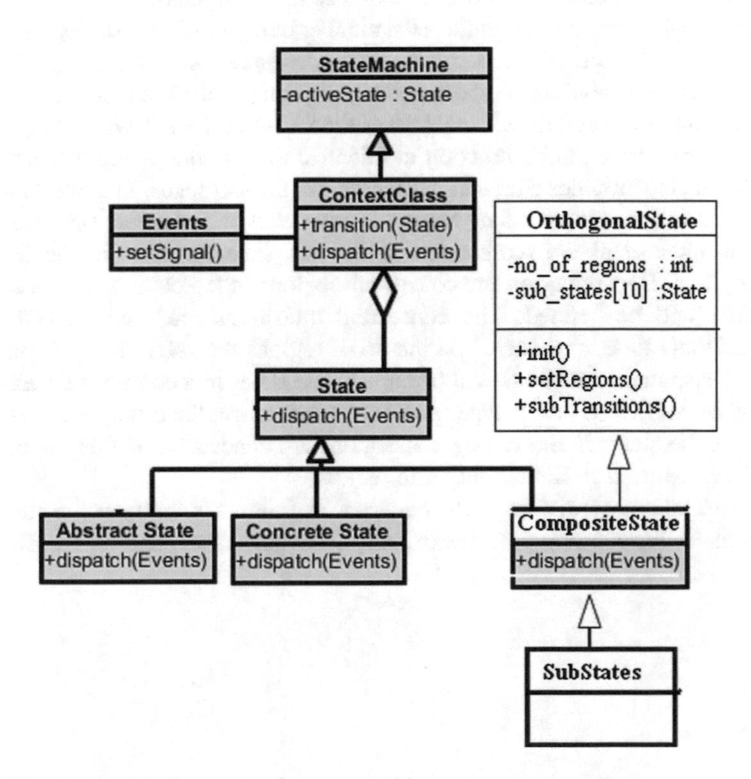

Fig. 4 The proposed design pattern for state machines

Table 1 Mapping state machine elements to oo program constructs

State machine element	Program construct
State	State class
Transition	Method in context class
Erem	Events class
Entry/exit actions	Method in state class
Internal action	Method in state class
Hierarchical states	Hierarchy of state classes
Concurrent transitions	Method in the orthogonalstate class

In the proposed method states and events are represented as classes. The state hierarchies are represented as hierarchy of State classes.

The state chart of a system will be represented as Context Class. It includes a state transition() function which changes the current state of the system. An event dispatch() function is also defined in the Context Class. It delegates the event handling to the respective State Classes.

State Class is the base class for deriving different state classes of the system. It defines an event dispatch function for supporting the function delegation. Each state in the system is derived as a child of State Class. In order to derive composite states, we define OrthogonalState Class in addition to the State Class.

The OrthogonalState Class defines the number of regions in the orthogonal states. In addition to that it records the active sub states in the regions. The number of active sub states varies depending on the number of regions. If there are concurrent states then the number of regions will be greater than 1. Otherwise it will be equal to 1. This class provides an init() function to initialize the number of regions and active sub states. It also provides a transition function for the sub states. The event handling is delegated to the corresponding State Classes based on the active sub states.

The parallel states are represented as the sub_state variables in the OrthogonalState Class. Depending on the concurrent regions in the composite state, the no of regions will be defined. The concurrent transitions are implemented as the subTransition() functions. Based on the no of regions the subTansition() functions and event dispatch() functions will be called. The states in a composite/hierarchical state can be AND type or OR type. AND type states form the orthogonal regions. It shows that the states in the orthogonal regions are concurrent. If there is only OR states, then we need to define only one region.

The skeletal code structure of the pattern is as follows. It includes the classes for Events, State, StateMachine, ContextClass, OrthogonalState, and CompositeState.

```
public class Events {
   public void setSignal(){
   }
}

public class State {
   public void dispatch(Bomb b, Events e){
   }
}

public class StateMachine{
    State activeState;
}

public class ContextClass extends StateMachine {
   public ContextClass () {
   }
   public void init(){
   }
   public void dispatch(Events e) {
   }
   public void tran(State target){
   }
}

public interface OrthogonalState {
      int no_of_regions;
      State[] sub_states;
   public void init();
   public void subTransition(int region, State target);
   public void setRegions(int no_of_regions );
}

public class CompositeState extends State implements OrthogonalState{
   public void init(){
   }
   public void subTransition(int region, State target) {
                  sub_states[i]=target;
   }
   public void dispatch(ConetxtClass context, Events e){
         sub_states[i].dispatch(context, e);
   }
   public void setRegions(int no_of_regions) {
          throw new UnsupportedOperationException("Not supported yet.");
   }
}
```

CompositeState class is inherited from State class as well as OrthogonalState class. In Java multiple inheritance is not directly supported. So, we make use of interface to define the OrthogonalState.

4 Case Study

We consider the case of an Explosion System which has four buttons to operate.

The system can be in *setting* state or *timing* state. In *setting* state we can set a defuse code, and increment or decrement the time out for the explosion system. In the *timing* mode the system exists in two concurrent states, *defuse* state and *tick* state.

The four events (corresponds to each button of the system) that can happen in the systems are, *UP*, *DOWN*, *ARM*, and *TICK*. When the system is in *setting* mode, if event *UP* occurs, it increments the *timeout* value. If the *DOWN* event occurs, then it decrements the *timeout* value. If the event *ARM* occurs, then the system changes to *timing* state. The *tick* event has no effect while the system is in *setting* mode.

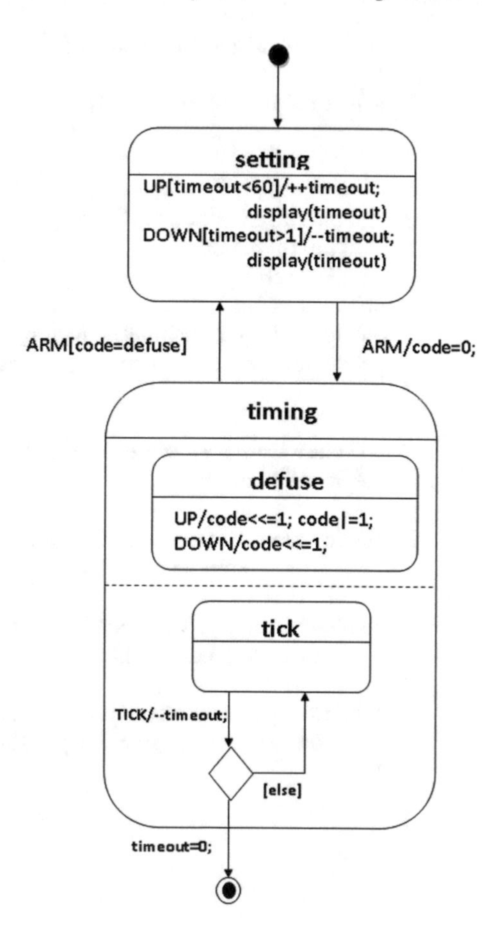

Fig. 5 UML state diagram representing the explosion system

Table 2 State transition table of the explosion system

Current state	Sub state	Events	[gurad]	Next state
Setting		UP	[timeout < 60]	Setting
		DOWN	timeout > 1	Setting
		ARM		Timing
		TICK		Setting
Timing	Diffusion	UP		Timing
		DOWN		Timing
		ARM	[code ==difuse	Setting
	Tick	TICK	[fine_time ==0]	Choice
			[timeout ==0]	Final
			[else]	Timing

When the system is in *timing* mode, the system will be in *defuse* state as well as *tick* state, two parallel states. If an *UP/DOWN* event occurs then it sets the secret key to defuse the system. When the *ARM* event occurs, it checks the secret key entered and the defuse code of the system. If it matches the system will be defused. In parallel, the *TICK* event exists where the time out decreases continuously. Each *TICK* event decreases the timeout value. When the time out reaches 0, the system explodes.

The state diagram of the explosion system is shown in Fig. 5 and the state transition table of the system is shown in Table 2. The implementation pattern of the Explosion system is shown in Fig. 6.

The context class here is the *ExplosionSystem.* It uses the Events class and the State class for setting the state of the system and event dispatching. The initial state is set to SettingState. Whenever an event encounters the corresponding event handling function will be called by using the run time polymorphism. Whenever the system enters the composite states, the init() function of the class has to be invoked. So this function call is included in the transition function.

```
public class ExplosionSystem extends StateMachine{
    ............
    ............
     public ExplosionSystem(byte defuse){
     m_defuse=defuse;
  }
  public void init(){
      m_timeout=10;
     tran(setting);
  }
  public void dispatch(Events e){
     m_state.dispatch(this, e);
  }
  public void tran(State target){
     m_state=target;
     if(m_state==timing)      {
        timing.init();
}    } }
```

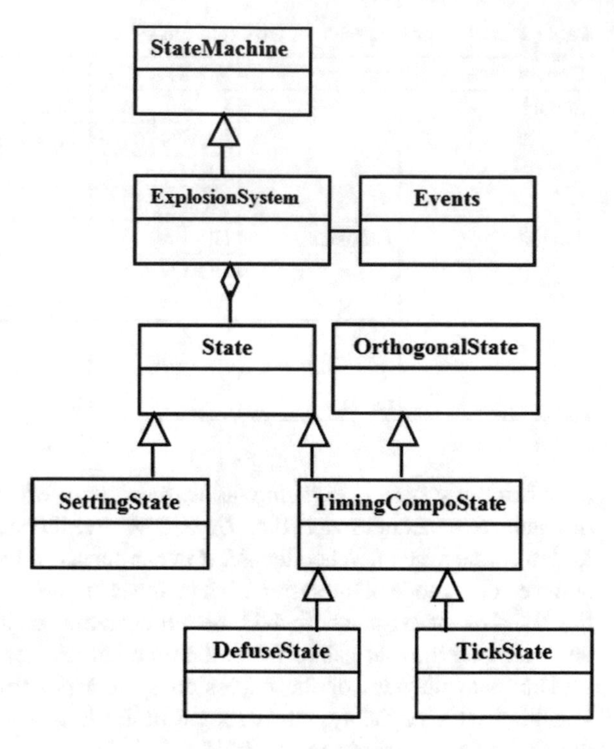

Fig. 6 The implementation pattern of the explosive system

The State class acts as the base class for deriving the states of the system. It consists of the event dispatch function which will be overridden in the child classes. The dispatch function takes the context class object and the event that occurred as the arguments. The overridden dispatch functions in the child classes will act as the event handlers. In different state classes the dispatch function is implemented differently according to how each state responds to that particular event.

```
public class State {
public void dispatch(ExplosionSystem b, Events e){
} }
```

The *setting* mode of the system is implemented as the SettingState which is derived from the State class. The dispatch function in this class handles the *UP, DOWN* and *ARM* events.

```
public class SettingState extends State {
  public void dispatch(ExplosionSystem context, Events e){
      switch(e.signal){
      case UP : {if(context.m_timeout<60){
               ++context.m_timeout;
               break;}}
      case DOWN :{if(context.m_timeout>1){
               --context.m_timeout;}
                 break;}
      case ARM  : {context.m_code=0;
                 context.tran(ExplosionSystem.timing);
                 break;}
}}}
```

The *timing* mode is a composite state which has two AND states, i.e., defuse state and tick state. So the timing mode is implemented as the composite state, TImingCompoState, which is derived from State class and OrthogonalState. The dispatch function delegates the events to the DefuseState and TickState. We store the active sub states and the sub transitions and event dispatching is done based on the active sub states.

```
public class TimingCompoState extends State implements OrthogonalState{
   .....................
   public void init()   {
      subTransition(1,tick);
      subTransition(2,diffuse);
   }
   public void subTransition(int region, State target)   {
      sub_states[region]=target;
   }
  public void dispatch(ExplosionSystem context, Events e){
      for(i=1;i<=no_of_regions;i++){
         sub_states[i].dispatch(context, e);
}   } }
```

The defuse state is implemented as the DefuseState class. It handles the *UP*, *DOWN* and *ARM* events. The tick state is implemented as the TickState class which handles the *TICK* event.

```
public class DefuseState extends TimingCompoState{
 public void dispatch(ExplosionSystem context, Events e){
    switch(e.signal) {
        case UP:context.m_code<<=1;
                 context.m_code|=1;
                 break;
        case DOWN:  context.m_code<<=1;
                    break;
        case ARM :if(context.m_code==context.m_defuse){
                 context.tran(ExplosionSystem.setting);
                 }else{
                System.out.println("Unsuccessful defusion");
                     }
                    break;
  } }}
public class TickState extends TimingCompoState{
 public void dispatch(ExplosionSystem context, Events e){
    switch(e.signal)
      { case TICK :  --context.m_timeout;
   if(context.m_timeout==0){                                    sm.blast(); }
 }  } }
```

The Events class defines the events that can happen in the explosive system.

```
public class Events {
.......................
   public void setSignal(int i){
      switch(i){
         case 1 : signal=events.UP;break;
         case 2 : signal=events.DOWN;break;
         case 3 : signal=events.ARM;break;
         case 4 : signal=events.TICK;break;
}    }}
```

5 Related Works

Dominguez et al. [4] presented a review of research works that propose methods to implement UML state chart diagrams. Dominguez et al. summarize their review by saying that the state transition process in most of the works are based on switch statement or state table. Another key finding of [4] is that very few papers support hierarchy and concurrency of states.

Ali [12] presents the implementation of concurrent and hierarchical state machines by making use of enumerators in Java language. It proposes a Java implementation pattern for state machines. The method presented in the paper is language dependent and so it cannot be extended to other languages like C++, C# etc.

Spinke [10] addresses concurrent and hierarchical state implementation. The paper proposed a double dispatch based event handling. The reaction of the state machine depends on the current state of the system as well as the event occurred. This is the theory behind double dispatch. The paper presents a case study to show case the proposed method. The implementation pattern presented in the paper is very bulky since it requires 17 classes in the implementation for representing a stat machine with 6 states and 6 events.

Niaz and Tanaka [8] propose a method to implement composite and concurrent states. In the proposed approach single event can trigger multiple transitions. This is against the semantics of the UML state machine. UML specifies that one event should be consumed for only one transition.

Schattkowsky [13] demonstrates how a fully featured UML 2.0 state machine can be represented using a small subset of the UML state machine features that enables efficient execution. They are trying to directly execute the state machines without converting it to implementation code. It is an alternative to native code generation approaches since it significantly increases portability. The paper describes the necessary model transformations in terms of graph transformations and discusses the underlying semantics and implications for execution.

Rudhal [14] presents a multi language code generator named as YAMDAT (Yet Another MDA Tool). As the name indicates, it's an MDA tool. It generates C++ and Java code from UML designs of the system. UML models will be represented in XML and this XML representation is the code model in the tool. They generate skeleton code for all methods and attributes in the UML class diagram. Moreover, unit test framework will be generated for the class. YAMDAT generates finite state machine class from each state diagram of the class.

6 Implementation and Evaluation

In this section we propose a code generator to generate prototype from UML State chart diagrams (SM) and Class Diagrams (CD). It takes the SM and CD in XML format as input. The Transformation Engine converts the SM and CD to Java source code. The transformation engine has three main parts, the Parsers, Transformation Rules and the Prototype Generator.

There are two parsers, one is CD Parser which parses the class diagram for code generation and the other one is SM Parser which parses the state machine. The Transformation Rule specifies which component in SM and CD converts to which programming construct. For example, a compound state has to be converted to an abstract super state class. The transformation rules strictly follow the design pattern proposed in this paper. It defines how to convert the compound states, parallel states and sequential states. This conversion rules are specified in Transformation Rules. The SM Prototype Generator converts the CD and SM to Java code by considering the Transformation Rules.

Table 3 Efficiency of SMConverter compared with Rhapsody and OCode

	OCode (ms)	Rhapsody (ms)	SMConverter (ms)	Efficiency over OCode	Efficiency over Rhapsody
Total time for events without transitions	4.4	5.05	4		
Average time per event without transition	0.00248	0.00284	0.00225	9.27	20.77
Total time for events having transitions	22.05	23.1	10.1		
Average time per event having transition	0.00992	0.01039	0.0045	54.64	56.69
Total time for all events	26.4	28.15	14.1		
Average time per event	0.0066	0.00704	0.00353	46.52	49.86

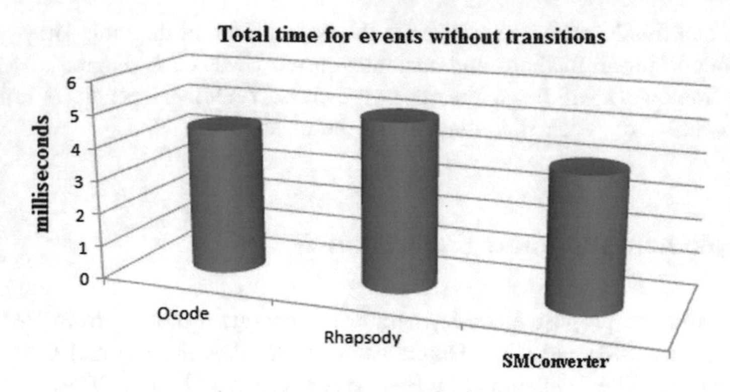

Fig. 7 Total time for events without transition

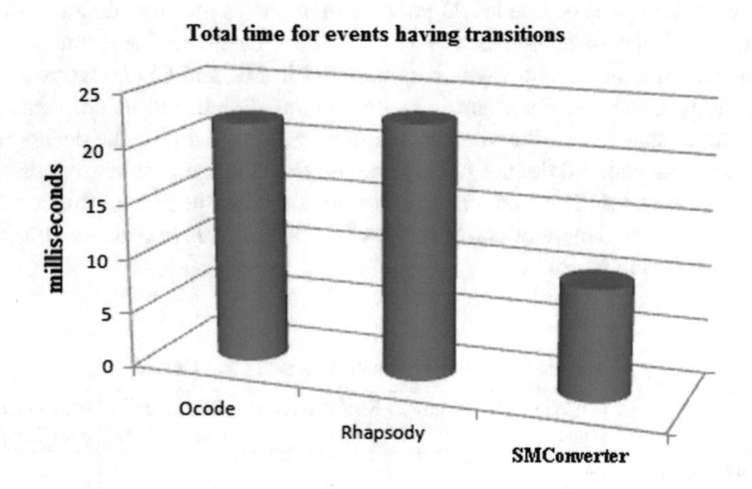

Fig. 8 Total time for events having transition

We have developed a tool, SMConverter, based on the proposed method. SMConverter is compared with similar tools like Rhapsody and OCode. The number of lines generated by each tool, the number of bytes generated and total number of classes generated is compared. Table 3 shows the comparison of SMConverter, Rhapsody and OCode. We considered the events with and without transitions. Total time taken for each type is calculated in milliseconds. Total number of requests for events without transition is 1778 and for events with transition is 2222. The efficiency of our tool (SMConverter) over other tools is shown in the Table 3. Figures 7 and 8 compare the total time taken for events without and with transition respectively.

7 Conclusion

Code generation from UML models is very essential in software development. In this paper, we introduce an object oriented method to implement the hierarchical and concurrent states in the state machine. The proposed approach helps us to implement both composition and concurrency with same design pattern. The case study and comparison with other tools reveals that the proposed approach gives less complex code and promising results. The object orientation that we used in the proposed approach provides flexibility in coding as well as its modification.

References

1. Rumbaugh, J., Jacobson, I., & Booch, G. (2007). *Object-oriented analysis and design with applications* (3rd ed.). Addison-Wesley.
2. Mellor, S. J., & Balcer, M. J. (2002) *Executable UML a foundation for model-driven architecture*. Addison-Wesley.
3. Ali, J., & Tanaka, J. (2001). Implementing the dynamic behavior represented as multiple state diagrams and activity diagrams. *Journal of Computer Science and Information Management (JCSIM), 2*(1), 24–34.
4. Dominguez, E., et al. (2012). A systematic review of code generation proposals from state machine specifications. *Journal of Information and Software Technology*, *54*, 1045–1066.
5. Douglass, B. P. (1998). *Real time UML—developing efficient objects for embedded systems*. Massachusetts: Addison-Wesley.
6. Jakimi, A., & Elkoutbi, M. (2009). Automatic code generation from UML state chart. *International Journal of Engineering and Technology, 1*(2), 165–168.
7. Ali, J., & Tanaka, J. (2000). Converting state charts into Java code. In *Proceedings Fourth World Conference on Integrated Design and Process Technology (IDPT'99)*, Dallas, Texas, USA.
8. Niaz, I. A., & Tanaka, J. (2005). An object-oriented approach to generate Java code from UML state charts. *International Journal of Computer and Information Science*, *6*(2).
9. Lazareviae, L., & Miliaev, D. (2000). Finite state machine automatic code generation. In *IASTED conference*, Austria.
10. Spinke, V. (2013) An object-oriented implementation of concurrent and hierarchical state machines, *Journal of Information and Software Technology*, *l–55*, 1726–1740.
11. Aabidi, M. H., et al. (2013). An object oriented approach to generate Java code from hierarchical-concurrent and history states. *International Journal of Information and Network Security*, *2*, 429–440.
12. Ali, J. (2010). Using Java Enums to implement concurrent-hierarchical state machines. *Journal of Software Engineering* *4*(3), 215–230. ISSN 1819-4311.
13. Schattkowsky, T., & Muller, W. (2005). Transformation of UML state machines for direct execution, VLHCC. In *Proceedings of the 2005 IEEE Symposium on Visual Languages and Human-Centric Computing* (pp. 117–124).
14. Rudhal, K. T., & Goldin, S. E. (2008). Adaptive multi-language code generation using YAMDAT. In *Proceedings of ECTI-CON 2008, Electrical Engineering/Electronics, Computer, Telecommunications and Information Technology, 5th International Conference, 2008,* 14–17 May 2008 (Vol. 1, pp. 181–184).

A New and Fast Variant of the Strict Strong Coloring Based Graph Distribution Algorithm

Nousseiba Guidoum, Meriem Bensouyad and Djamel-Eddine Saïdouni

Abstract We consider the state space explosion problem which is a fundamental obstacle in formal verification of critical systems. In this paper, we propose a fast algorithm for distributing state spaces on a network of workstations. Our solution is an improvement version of SSCGDA algorithm (for Strict Strong Coloring based Graph Distribution Algorithm) which introduced the coloring concept and dominance relation in graphs for finding the good distribution of given graphs [1]. We report on a thorough experimental study to evaluate the performance of this new algorithm. The quality of the proposed algorithm is illustrated by comparison with existing algorithms.

Keywords Formal verification · Graph distribution · Graph strict strong coloring · Combinatorial optimization · Heuristics · SSCGDA · GGSSCA

1 Introduction

Formal verification is one of the main approaches to aid engineers on the development and validation of concurrent systems. In this approach, system's behavior can be modeled as a state space. The set of states forms a graph where states are connected if there is an action that can be executed to transform the state into the other. The graphs, modeling finite state systems, can be explored exhaustively because of the graph size and the exploration time which may grow exponentially

N. Guidoum (✉) · M. Bensouyad · D.-E. Saïdouni
MISC Laboratory, A. Mehri, Constantine 2 University, Constantine, Algeria
e-mail: guidoum_nousseiba@hotmail.fr

M. Bensouyad
e-mail: meriem_bensouyad@hotmail.com

D.-E. Saïdouni
e-mail: saidounid@hotmail.com

R. Lee (ed.), *Software Engineering Research, Management and Applications*, Studies in Computational Intelligence 654,
DOI 10.1007/978-3-319-33903-0_11

with the size of the system description. In this case, the formal verification process becomes more and more slowly and may not be terminated. This problem is known as state space explosion problem [2–4]. One solution consists in the distribution of the graph. In fact, the need for parallel and distributed computing becomes inevitable to deal not only with the presented problem but with all applications using large scale graphs.

Graph distribution is a well-known optimization problem [5] where several factors should be taken into account to have a good distribution. The most important of them are the workload balancing of the workers (i.e. no unemployed or overloaded workers) and the minimization of the distribution cost (i.e. edges to be cut). In practice, graph distribution occurs in applications on many areas, others than formal verification, like parallel computing, communication protocols, distributed algorithms and industrial case studies.

Distributing the state space among several workstations (workers) which communicate through a message in a network is the subject of this paper. A heuristic based on coloring concept and dominance relation in graphs is presented to find a good distribution within a reasonable time.

Many papers present several approaches for solving this NP-hard problem [5] and other distributed applications on large graphs [6–8]. Authors in [9] have presented deeply the different solutions proposed up to 2009. All these solutions are based on a partition function which assigns each state to a fixed worker. These approaches differ mainly by the nature of this function. In [8], authors have used ten heuristics such that each one of them gives an algorithm for selecting the index of the partition where a state is assigned.

Recently, a new graph coloring based distribution approach is proposed [1, 10]. This approach is called "SSCGDA" (for Strict Strong Coloring based Graph Distribution Algorithm). It has been successfully employed to ensure a good distribution of a given graph modeling the system state space. The SSCGDA is the first heuristic algorithm, for resolving the state space distribution, based on the strict strong coloring. Particularly, the dominance property is used to make the initial distribution. After that, other processes are used to build the final distribution.

In this paper, the proposed approach adopts the same principle of SSCGDA algorithm to consolidate the idea of using the strict strong coloring for graph distribution but it runs significantly faster. The aim is to improve the algorithm performances and even its results quality. This is done by adjusting the initial distribution construction process and omitting some phases like grouping process.

The remainder of this paper is organized as follows: Sect. 2 gives some preliminary notations and definitions. In Sect. 3, the new proposed variant of the SSCGDA algorithm is presented. Section 4 discusses obtained experimental results. Conclusion and future works are given in the last section.

2 Preliminaries

Graphs are abstraction in computer science. The most common form of graph consists of a set of objects called vertices and set of edges.

State space representation of system's behavior is described as follow:

- All the states, the system can be in, are represented as vertices of a graph.
- A transition, that can change the system from one state to another, is represented by an edge from one node to another.

Let $G = (V, E)$ be a graph such that $V = \{v_1, v_2 \dots v_n\}$ is a finite set of vertices and $E \subseteq V \times V$ is a finite set of edges (nodes).

For any $u \in V$, let $N(u) = \{v \in V | (u, v) \in E\}$ be the set of adjacent vertices of u. Let $deg(u) = |N(u)|$ be the degree of the vertex u.

For $u \in V$ and $Y \subseteq V$, $u \sim Y$ means that u is adjacent to all vertices of Y. We say also u dominates the set Y.

2.1 Distribution Concept

The distribution problem has been formulated as an NP-complete graph optimization problem [5].

A distribution of state space graph on W network nodes (parts, workers) is a partition of its set of vertices into W pair-wise disjoint subsets ($V = \cup_{i=1}^{W} V_i$ and $V_i \cap V_j = \emptyset$ for all $1 \leq i, j \leq W$). We denote by E_{ij} the cross (external) edges (i.e. the set of edges between the vertices assigned to worker i and the vertices assigned to worker j). Then, the elements of the sets E_{ii} are internal (local) edges.

In fact, an efficient distribution should have:

- A minimal number of cross (external) edges E_{ij}, in other words, as many internal (local) edges E_{ii} as possible. We express this factor by the rate φ which is equal to the number of internal edges divided by the total number of edges:

$$\varphi(\%) = 100 \times \sum_{i=1}^{W} \frac{|E_{ii}|}{|E|}. \tag{1}$$

- Balance, i.e., more-or-less the same number of vertices on each part. To determine this, we establish the notion of standard deviation of the number of nodes, denoted by σ_V and defined as follows:

$$\sigma_V = \sqrt{\frac{1}{w}\sum_{i=1}^{w}(|V_i| - avg)^2}. \quad (2)$$

Such that, the average load *avg* is the total number of vertices divided by the parts number *W*: $avg = \frac{|V|}{W}$, since the vertices sets assigned to different parts are disjoints.

2.2 Coloring Concept

A proper vertex coloring of a graph is a vertex coloring such that no two adjacent vertices have the same color. A proper coloring *C* using at most *k* colors is called a proper *k*-coloring. It is a function $C: V \rightarrow \{1\ldots k\}$ such that $C(u) \neq C(v)$ for any $(u, v) \in E$ [11]. $C(v)$ is called the color of *v*. The vertices having the same color form a color class. Since each color class is an independent set of *G*, a coloring may also be seen as a partition of *V* into independent sets $(C_1, C_2 \ldots C_k)$ where $C_i = \{x \in V$ such that $C(x) = i\}$ [12]. A graph is *k*-colorable if it has a proper *k*-coloring. The chromatic number $\chi(G)$ is the minimum number of colors required for coloring *G*.

A strong coloring is a graph coloring such that for each vertex *v*, there is a color class C_i such that *v* is adjacent to every vertex of C_i (i.e. $u \sim C_i$). A graph is strongly *k*-colorable if it has a strong *k*-coloring [13].

A strict strong *k*-coloring (*k*-SSColoring) is a strong *k*-coloring such that there is non-empty color class. More formally, the *k*-SSColoring of *G* is a proper *k*-coloring $\{C_1, C_2 \ldots C_k\}$ of *G* such that for every vertex $u \in V$, there exists $i \in \{1, 2 \ldots k\}$ where *u* is adjacent to every vertex of C_i and $C_i \neq \emptyset$ (i.e. $u \sim C_i$). So, a strict strong coloring of *G* is a proper coloring of *G* such that every vertex of *G* is strict strong. We say that a color *i* is a dominated color by a vertex *u*, with $u \in V$ (or *u* dominates *i*) if and only if $u \sim C_i$ [14]. The strict strong chromatic number $\chi_{ss}(G)$ is defined as the minimum number of colors among all strict strong colorings [14].

3 Proposed Algorithm

3.1 Global Algorithm

The proposed algorithm is a refinement of the SSCGDA. It adopts the same principle where the purpose is to reduce as possible the algorithm complexity with preserving a good distribution.

The SSCGDA algorithm is divided into two phases: an initialization and an optimization processes. In the initialization process, it uses the strict strong coloring to make the initial distribution. Particularly, the first phase of the GGSSCA algorithm presented in [15] is used to give the initial number of parts and their centers. Then, it reorganizes the initial distribution using *the initial distribution construction process*. At the end of this step, each vertex colored with dominated color is considered as a center of each part and its neighbors which dominate this color class are put with it on the same part.

In the following process, *the grouping (splitting) process* is used to fusion (split) initial parts if the initial part number *DC* is upper (less) than the required number of parts *W*.

The basic idea of this new variant is to update the initialization step to give exactly (or less than) the available number of workers *W*. In this case, the GGSSCA algorithm should be updated.

Contrary to the SSCGDA algorithm which applies the first step of the GGSSCA until obtaining *DC* centers, the new variant uses a modified GGSSCA algorithm to get exactly (or less than) *W* centers. In this case, the grouping process used in SSCGDA, will be useless in this version. This minimizes the running time and reduces the algorithm complexity. Consequently, the proposed algorithm will be more simple and faster than the SSCGDA.

Box 1. The Global Algorithm.

```
Input: Graph G= (V, E).
W: the number of available workers.
Output:πw distribution of graph G into W sub-graphs
         (parts).
Initialization:
 Centers: the set of colored vertices with dominated
    colors, obtained after applying the modified GGSSCA.
Begin
 Centers := Modified GGSSCA();
 Initial distribution construction process();
 If (|Centers| < W) Then
     Splitting process();
 End if
End
```

In the case where the dominated colors number obtained after applying the modified GGSSCA algorithm is less than *W*, the splitting process will be used after the construction of the initial distribution to split the obtained parts.

3.2 Modified Generalized Graph Strict Strong Coloring Algorithm

Box 2. The Modified GGSSCA Algorithm.

```
Input: A graph G = (V, E).
       W:  the number of available workers
Output: Centers.
Initialization:
   Satisfied: the set of vertices having the dominance
         property. Initially this variable is empty.
   Colored: the set of colored vertices. Initially this
         variable is empty.
   D ⊆ V × Integer: Is a set containing a degree of vertex
         at a given calculation step.
Begin
  Repeat
    D:= ∪u∈(V−Colored) {(u, deg(u) − |N(u) ∩ Satisfied|};
    Select (u,n) such that u has maximal number n in D;
    Colored := Colored ∪ {u};
    Satisfied := Satisfied ∪ N(u);
  Until ((|Colored| = W) or (Satisfied=V))
  Centers = Colored;
End
```

The GGSSCA algorithm was used to make the initial distribution in the SSCGDA algorithm. This latter has used the first phase of GGSSCA which ensures the dominance property. At the end of this phase, each vertex is adjacent to at least one non empty color class and the dominated colors number *DC* is given to define the number of initial parts.

However, in this paper, this heuristic is updated to obtain exactly (or less than) *W* parts centers (see Box 2). The goal is to reduce the number of iterations. This implies the minimization of the process complexity: the complexity of the GGSSCA based process used in the basic version of SSCGDA algorithm is $O(|V|^2)$ while the complexity of the updated one, given in Box 2, is only $O(W \times |V|)$.

3.3 Initial Distribution Construction

```
Box 3. The initial distribution construction process.
Input: Graph G= (V, E).
       Centers: the set of colored vertices with domi-
                nated colors, obtained after applying
                the modified GGSSCA, considered as
                parts centers.
Output: Partition Π|Centers| of G.
Initialization:
  Π|Centers| = {V1, V2,… V|Centers|}: such that Vi ∈ Π|Centers| is
   an independent part. Initially Vi ={u|u ∈ Centers}.
  Not_Conflicting_verticesu: the neighbors set of u
        which dominates only the color class contain-
        ing the vertex u.
  Conflicting _vertices: the set of vertices that dom-
        inate more than one color class.
  Not_Adjacent: the set of vertices that do not domi-
        nate any color.
Begin
 For (All parts Vi) Do
  Attribute all vertices of Not_Conflicting _verticesu
  set to the appropriate part containing u as center;
 End For
 For (All x ∈ Conflicting _vertices) Do
  Attribute each vertex x to the appropriate part Vi
  where x has the maximum connections with Vi;
 End For
 For (All x' ∈ Conflicting _vertices) Do
  Attribute each vertex x' to the appropriate part Vj
  where x' has the maximum connections with Vj;
 End For
End
```

The initial distribution construction process (See Box 3) is used for constructing the initial pair-wise disjoint parts.

After applying the updated GGSSCA algorithm, we get centers for all parts such that the part center is defined as a colored vertex with dominated color.

Each part V_i initially contains exactly one center which is a colored vertex ($u \in$ *Centers*). Then, its neighbors (*Not_Conflicting _vertices*$_u$ set), that dominate only the color class containing the colored vertex u, will be added to V_i.

For the vertices (*Conflicting _vertices* set), that dominate more than one color class, each one will be added to the appropriate part V_j having with it the maximum connections.

After associating all vertices that dominate at least one color, it remains the vertices that are not adjacent to any center. For that, each vertex of them will be added to the part that has with it maximum connections.

3.4 Splitting Process

The splitting process (see Box 4) selects, at each iteration, the part V_i having the maximum number of vertices for obtaining a relaxed balanced between different parts. Then, it searches in V_i the vertices which have a strong connection in order to keep them in the new and same part $V_{|\pi+1|}$. Finally, the distribution will be updated by adding the new part $V_{|\pi+1|}$ and removing all vertices existing in it from the main part V_i. This process will be repeated until obtaining W parts.

Box 4. The splitting process.

```
Input: Π|Centers| initial distribution of G.
       W: number of workers.
Output: Πw distribution of G with W parts.
Begin
  Π := Π|Centers|;
  Repeat
      Select Vi which has the maximum number of nodes;
      Compute the connectivity between nodes in Vi;
      Put the vertices, strongly connected, in the par-
      tition V|π|+1;
      Remove all vertices existing in V|π|+1from Vi ;
  Until (|Π |= W)
  Πw = Π;
End
```

4 Experiments

To show the quality of the proposed algorithm, a comparison with existing approaches is conducted. For instance, we have picked the basic approach SSCGDA [1] and the hash function (MD5) based algorithm [6]. In fact, the MD5 based algorithm offers a distributed state space, using a hash function based on the function of encoding MD5 (for Message Digest Algorithm 5).

In order to illustrate the performance of the proposed approach, we consider measurements given above (1 and 2). The experiment consists of the distributing state space of the systems and the computation of given measurements.

In the context of our experiments, we have selected 3 well known classic case studies in system models. These models include dining philosophers system [16], Peterson solution for mutual exclusion [17] and shared memory system [18].

The obtained results are shown in Tables 1 and 2 where:

- $|V|$ denotes the number of vertices,
- $|E|$ denotes the number of edges,
- W is the required number of parts (workers),
- σ_V is the standard deviation of vertices on each part,
  It is important to note that the standard derivation, reported in this paper, is calculated after removing the outlier points. Outliers are extreme (exceptional) values that stand out from the other values of a data set. If not removed, these extreme values can skew the conclusions that might be drawn from the data in question. To identify and remove these outlying values, a statistical test, called Grubbs [19], has been applied on our data set (the workload of each part of the partition).

- φ: the fraction of the local connections number divided by the number of all connections.
- *DC* is the number of dominated colors obtained by the first phase of GGSSCA,
- *T-exe*: the average execution run-time (in seconds) is taken using 10 runs.

Table 1 Comparative results of the proposed approach, SSCGDA and MD5 based algorithm

W = 20	|V|	|E|	DC	Moy	Approach	σ_V	φ (%)
Philosophers	729	3402	127	36	MD5 based algorithm	18.92	6
					SSCGDA algorithm	11.04	46.2
					Proposed algorithm	7.9	52.94
Shared memory	8019	52974	432	400	MD5 based algorithm	201.26	7
					SSCGDA algorithm	176.70	65.7
					Proposed algorithm	48.13	56.88
Peterson	20754	62262	4669	1037	MD5 based algorithm	519	7
					SSCGDA algorithm	387.80	91.2
					Proposed algorithm	1434.4	90.9

Table 2 Comparative results of the proposed approach and SSCGDA algorithm relative to execution time

W = 20	|V|	|E|	DC	T-exe SSCGDA	T-exe Proposed approach
Philosophers	729	3402	127	0.27	0.24
Shared memory	8019	2974	432	17.32	11.16
Peterson	20754	2262	4669	1251.82	55.77

According to the results reported on Table 1, we can remark that, in all cases, the proposed approach gives a quite good rate of local edges and provides a good balance.

Compared to the hash function MD5 based algorithm, the proposed one gives a significant improvement in both balance and local edges rate. For instance, the improvement reaches 83 % of local connections rate for the Peterson graph.

Compared to the SSCGDA algorithm, it is remarked that the results vary according to the experimental model:

For philosophers' graph, the proposed approach gives the well results in term of both balance and local edges rate compared to the ones given by the basic version of SSCGDA.

For shared memory graph, although the proposed approach does not give a better local edges rate compared to SSCGDA but it can guarantee the best load balancing. This means that the network workers are well balanced when using the proposed approach.

For the last model: Peterson, the proposed approach does not provide the better results compared to SSCGDA.

The diversity of the obtained results can be explained as follow. In fact, the quality of the proposed approach depends enormously on the behavior of the systems and the choice of the number of workers. After several runs, we have observed that the proposed algorithm gives good results when the number of workers W is near to the number of dominated colors DC. This can be explained through the example of Peterson: By applying SSCGDA, 4669 centers are initially obtained which are well distributed among all the graph contrary to the proposed algorithm which manipulates only 20 centers. This limited number cannot be well distributed on the entirely graph. For that, we have obtained this balance interval.

Regarding now the running time, it appears that the basic version of SSCGDA algorithm takes more time compared to the proposed one (see Table 2). Indeed, we observe that it depends on the number of vertices, edges and also the number of workers.

In addition to keeping good results, the brought modifications in this approach are adjusted to achieve a significant improvement in term of execution time. This improvement is outlined due to:

- The update of GGSSCA where its complexity is decreased. It becomes $O(W \times |V|)$ instead $O(|V|^2)$.
- The omission of the grouping process which affects enormously the execution time.

According to the Table 2, the reader can remark that the proposed approach is fast and efficient in term of execution time compared to the basic version of SSCGDA algorithm.

5 Conclusion

In this paper, we have presented a new approach based on graph coloring algorithm to solve the graph distribution problem. This approach is based on the graph strict strong coloring concept. More precisely, we have used the modified GGSSCA algorithm which ensures the dominance property. This latter is exploited to initially distribute the graph and obtain W or less pair-wise disjoint parts. In the case where the obtained number of dominated colors DC is less than the required number W, the splitting process is used to find a good distribution with W parts.

The algorithm is experimented on several graphs with different nature. The obtained results showed that the presented heuristic gives generally good statistics compared to the SSCGDA and hash function MD5 based algorithm. In addition, we have observed that the performance quality depends on the choice of the number of workers (W): It gives an optimal distribution when the number of workers is closer to the number of dominated colors. This property will be very useful in future works.

To put in practice the result of this work, the proposed approach will be improved by integrating the distribution process during the graph generation. Also, it remains to see the effect of the proposed method on the performance of verification algorithms.

References

1. Guidoum, N., Bensouyad, M., & Saïdouni, D. E. (2013). The strict strong coloring based graph distribution algorithm. *International Journal of Applied Metaheuristic Computing, 4*, 50–66.
2. Valmari, A. (1998). The state explosion problem. In *Lectures on Petri Nets I: Basic Models: Of Lecture Notes in Computer Science* (vol. 1491, pp. 429–528). London, UK.
3. Clarke, E., Grumberg, O., & Peled, D. (1999). *Model checking*. Cambridge, MA: The MIT Press.
4. Bérard, B., Bidoit, M., Finkel, A., Laroussinie, F., Petit, A., Petrucci, L., et al. (2001). *Systems and software verification: Model-checking techniques and tools*. Springer.
5. Bixby, R., Kennedy, K., & Kremer, U. (1993). *Automatic data layout using 0–1 integer programming*, Houston, TX, United State: Rice University, Center for Research on Parallel Computation, Tech. Rep. CRPC-TR93349-S.
6. Bouneb, Z., & Saïdouni, D. E. (2009). Parallel state space construction for a model checking based on maximality semantics. In *Proceedings of the 2nd Mediterranean Conference on Intelligent Systems and Automation* (vol. 1107, pp. 7–12).
7. Orzan, S., van de Pol, S., & Valero Espada, M. (2005). A state space distribution policy based on abstract interpretation. *Electronic Notes in Theoretical Computer Science, 128*(3), 35–45.
8. Stanton, I., & Kliot, G. (2011). *Streaming graph partitioning for large distributed graphs* (Tech. Rep. MSR-TR-2011-121). Microsoft Research Lab.
9. Saad, R., Dal Zilio, S., Berthomieu, B,. & Vernadat, F. (2009). Enumerative parallel and distributed state space construction. In *Ecoled'Eté Temps Réel (ETR'09)*, Paris, France.

10. Bensouyad, M., Bouzenada, M., Guidoum, N., & Saïdouni, D. E. (2014). A generalized graph strict strong coloring algorithm: Application on graph distribution. *Contemporary Advancements in Information Technology Development in Dynamic Environments*, 181.
11. Klotz, W. (2002). *Graph coloring algorithms*. Clausthal, Germany: Clausthal University of Technology, Tech. Rep. No.
12. Dharwadker, A. (2006). *The independent set algorithm*. Institute of Mathematics. Retrieved from http://www.dharwadker.org/independent_set/.
13. ZverovichI, E. (2006). A new kind of graph coloring. *Journal of Algorithms*, *58*(2), 118–133.
14. Haddad, M., & Kheddouci, H. (2009). A strict strong coloring of trees. *Information Processing Letters, 109*(18), 1047–1054.
15. Bouzenada, M., Bensouyad, M., Guidoum, N., Reghioua, A., & Saïdouni, D. E. (2012). A generalized graph strict strong coloring algorithm. *International Journal of Applied Metaheuristic Computing (IJAMC)*, *3*(1), 24–33.
16. NetLogo Models Library: Sample Models/Computer Science Standards, ccl.northwestern.edu. Retrieved from http://ccl.northwestern.edu/netlogo/models/DiningPhilosophers.
17. Model Checking Contest, "Peterson" model, sumo.lip6.fr. Retrieved from http://sumo.lip6.fr/Peterson_model.html.
18. Dijkstra, E. W. (1965). Solution of a problem in concurrent programming control. *CACM, 8*(9), 569. doi:10.1145/365559.365617.
19. Grubbs, F. E. (1969). Procedures for detecting outlying observations in samples. *Technometrics, 11*(1), 1–21.

High Level Petri Net Modelling and Analysis of Flexible Web Services Composition

Ahmed Kheldoun, Kamel Barkaoui, Malika Ioualalen and Djaouida Dahmani

Abstract In this paper we propose a model to deal with flexibility in complex Web services composition (WSC). In this context, we use a model based on high level Petri nets called RECATNets, where control and data flows are easily supported. Indeed, RECATNets combine the strengths of recursive Petri nets with the expressive power of abstract data types. Since RECATNets semantics is expressed in terms of the conditional rewriting logic, one can use the Maude LTL Model-Checker to investigate several behavioral properties of Web services composition.

1 Introduction

With the increasing complexity of business requirements, the distributed and flexible characteristics of Web services, the possibility of errors in Web service composition (WSC for short) is greatly increased. As a result, many researchers tried to propose formal methods, Finite State Machine [1], Pi calculus [2] or Petri nets [3, 4] to build the formal description and verification models of WSC. However, one of the weaknesses of these methods is their lack of support for managing flexible WSC which require dynamic adaptation of their structure. We refer to flexible WSC as the ability to create, modify, extend or suppress (sub)processes in a structured way, at the

A. Kheldoun (✉) · M. Ioualalen · D. Dahmani
MOVEP, Computer Science Department, USTHB, Algiers, Algeria
e-mail: ahmedkheldoun@yahoo.fr

M. Ioualalen
e-mail: mioualalen@usthb.dz

D. Dahmani
e-mail: ddahmani2000@yahoo.com

A. Kheldoun
Sciences and Technology Faculty, Yahia Fares University, Medea, Algeria

K. Barkaoui
CEDRIC-CNAM, 292 Rue Saint-Martin, 75141 Paris Cedex 03, France
e-mail: kamel.barkaoui@cnam.fr

R. Lee (ed.), *Software Engineering Research, Management and Applications*, Studies in Computational Intelligence 654,
DOI 10.1007/978-3-319-33903-0_12

occurrence of exceptions. In addition, such a composite Web services can potentially be very large, complex and cumbersome. In regard to the previous points, if we want to describe, faithfully, real-life WSC, we need an expressive modeling formalism that allows, in one hand, to specify their flexible and distributed features, and in other hand, to check the interaction (control-flow) correctness of these business processes while taking into account their data flow aspect. In this paper, a new model based on a kind of high level algebraic Petri nets combining the expressive power of abstract data types and Recursive Petri nets [5] called Recursive ECATNets (RECATNets for short) [6] is proposed in order to cope with the flexibility problem in complex WSC. The RECATNets model offers practical mechanisms for a direct and intuitive support of dynamic creation and suppression of processes. They are well-suited for handling the most advanced WSC patterns (involving cancellation and multiple instances). The proposed model is expressive enough to capture the semantics of complex service compositions and their respective specificities. Since RECATNets semantics is expressed in terms of the conditional rewriting logic [7], one can use the Maude LTL model-checker [8] to investigate several behavioral properties of Web services composition. The remainder of this paper is organized as follows. Section 2 gives a brief overview of related work. Section 3 presents the basic concepts of RECTANets. Web service modeling and specification using RECATNet are presented in Sect. 4. Section 5 is devoted to the algebra for composing Web services and its RECATNets-based formal semantics. A case study is presented in Sect. 6. Section 7 presents the analysis method and the verification process. Finally, Sect. 8 concludes and gives some further research directions.

2 Related Works

The composition of web services requires the modelling of different combinations of web services involved in this composition [9]. The modelling of web services composition is addressed in several papers. In this section, we briefly overview some approaches that are closely related to our work. In [10], the authors propose in their project e-flow to use workflow management system in order to compose web services. However, this approach lacks a formal model for specifying web services composition. In [11], the authors developed a Petri net based approach that uses several structural properties for identifying inconsistent dependency specification in a workflow. However, the proposed approach is restricted to acyclic workflows. In [3], the authors propose a Petri net-based algebra for composing web services. They provide a direct mapping from each composition operator to Petri nets. Their model is expressive enough; but data types cannot distinguish because they used elementary Petri nets. Contrary to our model, data types can be distinguished because the model used the expressive power of abstract data types. In [4], the authors used colored Petri nets [12] for modelling web services and their composition where data types can be distinguished. However, the author focalise in modeling, only, simple patterns. The author propose in [13] a model for composing web services based high level Petri nets, called G-nets. In this model, the authors propose to compose web services via special places called instantiated switch places. For the analysis, the author need to tranform their G-net models into Predicate/Transition nets (PrT-nets). However, some useful patterns like multiple instance and cancellation of service are not addressed. In [14], the author present a review of forty-three patterns for modeling business process using Colored Petri-Net (CPN). However, we note that the pattern of multiple instantiation of a sub-process is difficult to implement when a particular instance of a sub-process initiates other sub-process instances or involves recursive calls to the one of these ancestors process. This is even more complex when the number of such instances is not known prior to the execution of the process or where such instances require synchronization on many levels. One of the weakness of the previous approach is their lack a support for modeling useful advanced patterns like multiple instance and cancellation of service. In order to address this issue, we present a modular and hierarchical formalism called RECATNet that allows composition via abstract transitions. The usefulness of our proposed model is: (1) offering a practical mechanisms for handling the most advanced flow patterns (dynamic) multiple instance and cancellation of Web service, (2) providing a hierarchical composition of web services, (3) its modular specification and its flexibility by adding/removing service's instances in a dynamic manner, (4) allowing distributed execution of web services composition and (5) its semantic may be defined in terms of conditional rewriting logic [7] therefore, the model-checker MAUDE [8] can be used to check its correctness.

3 Recursive ECATNet Review

Recursive ECATNets (abbreviated RECATNets) [6] are a kind of high level algebraic Petri nets combining the expressive power of abstract data types and Recursive Petri nets [5]. Each place in such a net is associated to a sort (i.e. a data type of the underlying algebraic specification associated to this net). The marking of a place is a multiset of algebraic terms (without variables) of the same sort of this place. Moreover, transitions in RECATNet are partitioned into two types (Fig. 1): elementary and abstract transitions. Each abstract transition is associated to a starting marking, denoted like a multi-set of places put inside bracket. A capacity associated to a place p specifies the number of algebraic terms which can be contained in this place for

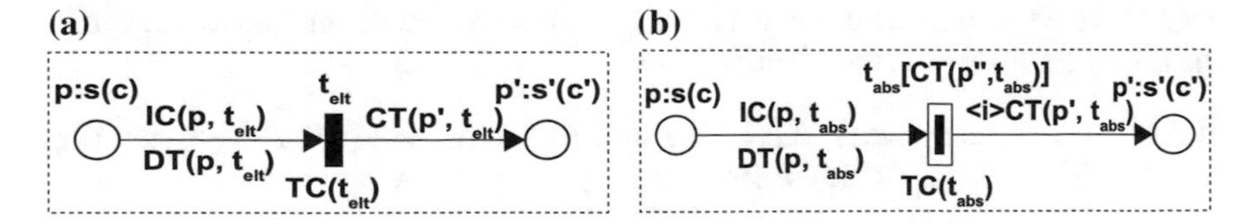

Fig. 1 Transition types in RECATNets. **a** Elementary transition. **b** Abstract transition

Table 1 Different forms of Input Condition IC(p, t)

IC(p, t)	Enabling condition
a^0	The marking of the place p must be equal to a (e.g. $IC(p,t) = \phi^0$ means the marking of p must be empty)
a^+	The marking of the place p must include a (e.g. $IC(p,t) = \phi^+$ means condition is always satisfied)
a^-	The marking of the place p must not include a, with $a \neq \phi$
$\alpha 1 \wedge \alpha 2$	Conditions $\alpha 1$ and $\alpha 2$ are both true
$\alpha 1 \vee \alpha 2$	$\alpha 1$ or $\alpha 2$ is true

each element of the sort associated to p. As shown in Fig. 1, the places p and p' are respectively associated to the sorts s and s' and to the capacity c and c'. An arc from an input place p to a transition t (elementary or abstract) is labelled by two algebraic expressions $IC(p,t)$ (Input Condition) and $DT(p,t)$ (Destroyed Tokens). The expression $IC(p,t)$ specifies the partial condition on the marking of the place p for the enabling of t (see Table 1). The expression $DT(p,t)$ specifies the multiset of terms to be removed from the marking of place p when t is fired. Also, each transition t may be labelled by a Boolean expression $TC(t)$ which specifies an additional enabling condition on the values taken by contextual variables of t (i.e. local variables of the expressions IC and DT labelling all the input arcs of t). When the condition $TC(t)$ is omitted, the default value is the term *True*. For an elementary transition t, an output arc (t,p') connecting this transition t to a place p' is labelled by the expression $CT(t,p')$ (*Created Tokens*). However, for an abstract transition t, an output arc (t,p') is labelled by the expression $\langle i \rangle CT(t,p')$ (*Indexed Created Tokens*). These two algebraic expressions specify the multiset of terms to produce in the output place p' when the transition t is fired. In the graphical representation of RECATNets, we note the capacity of a place regarding an element of its sort only if this number is finite. If $IC(p,t) =_{def} DT(p,t)$ on input arc (p,t) (e.g. $IC(p,t) = a^+$ and $DT(p,t) = a$), the expression $DT(p,t)$ is omitted on this arc. In what follows, we note $Spec = (\Sigma, E)$ an algebraic specification of an abstract data type associated to a RECATNet, where $\Sigma = (S, OP)$ is its multi-sort signature (S is a finite set of sort symbols and OP is a finite set operations, such $OP \cap S = \phi$). E is the set of equations associated to *Spec*. $X = (X_s)_{s \in S}$ is a set of disjoint variables associated to *Spec* where $OP \cap X = \phi$ and X_s is the set of variables of sort s. We denote by $T_{\Sigma,s}(X)$ the set of S-sorted S-terms with variables in the set $X.[T_\Sigma(X)]_\oplus$ denotes the set of the multisets of the Σ-terms $T_\Sigma(X)$ where the multiset union operator $(_\oplus)$ is associative, commutative and admits the empty multiset ϕ as the identity element.

Definition 1 A recursive ECATNet is a tuple $RECATNet = \langle Spec; P, T, F; sort, Cap, IC, DT, CT, TC, I, \Upsilon, ICT, K \rangle$ where:

- $Spec = (\Sigma, E)$ is a many sorted algebra where the sorts domains are finite (with $\Sigma = (S, OP)$), and $X = (X_s)_{s \in S}$ is a set of S-sorted variables,

- $[P, T, F]$ is a net where ($\text{T} \cap \text{P} = \phi$) and $T = T_{elt} \cup T_{abs}$ is finite set of transitions partitioned into abstract and elementary ones. T_{abs} and T_{elt} denoted the set of abstract and elementary transitions,
- sort: $P \to S$, is a mapping called a sort assignment,
- *Cap*: is a P-vector on capacity places: p∈P, Cap(p): $T_{\Sigma}(\phi) \to \mathbb{N} \cup \{\infty\}$,
- $IC : P \times T \to [T_{\Sigma,sort(p)}(X)]^{*}_{\oplus}$ where $* \in \{0, +, -\}$ maps a multiset of terms for every input arc,
- $DT : P \times T \to [T_{\Sigma,sort(p)}(X)]_{\oplus}$ maps a multiset of terms for every input arc,
- $CT : P \times T \to [T_{\Sigma,sort(p)}(X)]_{\oplus}$ maps a multiset of terms for every output arc (p, t) where $t \in T_{elt}$ and a starting marking associated to $t \in T_{abs}$ according to place p,
- $TC : T \to [T_{\Sigma,bool}(X)]$ maps a boolean expression for each transition,
- $I = I_{cut} \cup I_{pre}$ is a finite set of indices, called termination indices, dedicated to cut steps and preemptions (interruptions) respectively,
- Υ is a family, indexed by I, of effective representation of semi-linear sets of final markings,
- $ICT : P \times T_{abs} \times I \to [T_{\Sigma,sort(p)}(X)]_{\oplus}$ maps a multiset of terms for every output arc (p, t, i) where $t \in T_{abs}$ and $i \in I$,
- $K : T_{elt} \to T_{abs} \times I_{pre}$, maps a set of interrupted abstract trasitions, and their associated termination indexes, for every elementary transition.

Let's use the net presented in Fig. 2a to highlight RECATNet's graphical symbols and associated notations. (1) An elementary transition is represented by a filled rectangle; its name is possibly followed by a set of terms $t'\langle i \rangle \in T_{abs} \times I$. Each term specifies an abstract transition t' which is under the control of t, associated with a termination index to be used when aborting t' consequently to a firing of t. For instance t_{cancel} is an elementary transition where its firing preempts threads started by the firing of t_1 and the associated termination index is *1*. (2) An abstract transition t is represented by a double rectangles with the center filled; its name is followed by the starting marking $CT(t)$. For instance, t_1 is an abstract transition and $CT(t_1) = \langle p_5, Rq \rangle$ means that any thread created by firing of t_1 starts with one token i.e. one request Rq in place p_5. (3) Any termination set can be defined concisely based on place marking. For instance, Υ_0 specifies the final marking of threads such that the place p_6 is marked at least by one token. (4) The set I of termination indices is deduced from the indices used to subscript the termination sets and from the indices bound to elementary

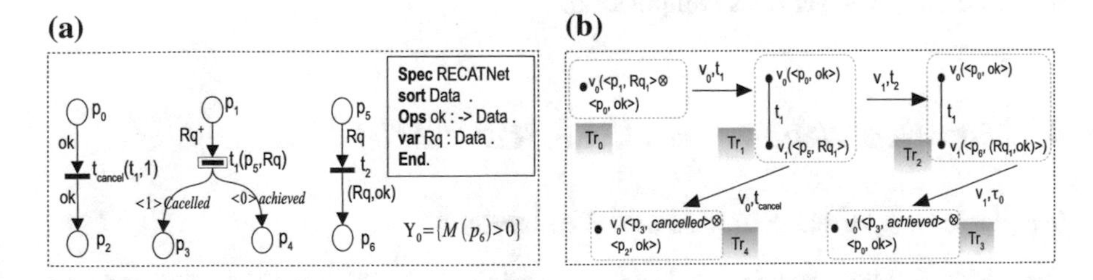

Fig. 2 Example of a RECATNet and two possible firing sequences

transitions i.e. interruption. From the example $I = \{0, 1\}$. Informally, a RECATNet generates during its execution a dynamical tree of marked threads called an extended marking, which reflects the global state of such net. This latter denotes the fatherhood relation between the generated threads (describing the inter-threads calls). Each of these threads has its own execution context.

Definition 2 (*Extended marking*) An extended marking of a RECATNet is a labelled rooted tree denoted $Tr = \langle V, M, E, A \rangle$ where:

- V is the set of nodes (i.e. threads),
- M is a Mapping $V \rightarrow [T_{\Sigma}(\phi)]_{\oplus}$ associating an ordinary marking with each node of the tree, such that $\forall v \in V, \forall p \in P, M(v)(p) \leq Cap(p)$,
- $E \subseteq V \times V$ is the set of edges,
- A is a mapping $E \rightarrow T_{abs}$ associating an abstract transition with each edge.

Note that contrary to ordinary nets, RECATNet are often disconnected since each connected component may be activated by the firing of abstract transitions.

Running example. Figure 2b highlights a possible firing sequences of the RECATNet represented in Fig. 2a. The graphical representation of any extended marking Tr is a tree where an arc $v_1(m_1) \rightarrow v_2(m_2)$ labeled by t_{abs} means that v_2 is a child of v_1 created by firing abstract transition t_{abs} and m_1 (resp. m_2) is the marking of v_1 (resp. v_2). Note that the initial extended marking Tr_0 is reduced to a single node v_0 whose marking is $\langle p_1, Rq_1 \rangle \otimes \langle p_0, ok \rangle$. From the initial extended marking Tr_0, the abstract transition t_1 is enabled; its firing leads to the extended marking Tr_1 which contains a fresh node v_1 marked by the starting marking $CT(t_1)$. Then, the firing of the elementary transition t_2 from node v_1 of Tr_1 leads to an extended marking Tr_2, having the same structure as Tr_1 but only the marking of node v_1 is changed. From node v_1 in Tr_2, the cut step τ_0 is enabled; its firing leads to an extended marking Tr_3 by removing the node v_1 and change the marking on its node father i.e. v_0 by adding $ICT(t_1, 0) = (p_4, achieved)$. Also, another way to remove nodes in extended marking using elementary transition with associated preemption. For instance, from node v_0 in Tr_1, the elementary transition t_{cancel} with associated preemption $(t_1, 1)$ is enabled; its firing leads to an extended marking Tr_4 by removing the node v_1 and change the marking on its node father i.e. v_0 by adding $ICT(t_1, 1) = (p_3, cancelled)$. More details about RECATNets such as formal firing and fundamental properties are presented in [6, 15]. This paper shows the usefulness of using the formalism of RECATNets in the field of Web services composition.

4 Modeling Web Services Using RECATNet

We give now a formal definition of a Web service.

Definition 3 (*Web Service*) A Web service is a tuple [3] $S = \langle NameS, Desc, Loc, URL, CS, RECATNetS \rangle$ where:

- *NameS* is the name of the service used as its unique identifier,
- *Desc* is the description of the provided service. It summarizes what functionalities the service offers,
- *Loc* is the server in which the service is located,
- *URL* is the invocation of the Web service,
- *CS* is a set of the component services of the Web service, if $CS = \{NameS\}$ then S is a basic service, otherwise$S$ is a Composite service,
- $RECATNetS = \langle Spec; P, T, F; sort, Cap, IC, DT, CT, TC, I, \Upsilon, ICT, K\rangle$ is the RECATNet service modeling the dynamic behavior of the Web service.

We show in the next section how Web services can be incrementally composed.

5 Web Services Composition

The common control structure in Web services composition usually includes simple patterns like: sequential, choice, iteration, parallel and discriminator operators and complex patterns like multiple instance and cancellation service [14]. Suppose S_1, S_2 and S_3 are three different atomic Web services i.e. each service S_i perfoms an individual operation that cannot be split into sub-operations. The algebra operator descriptions of these control structure can be seen as: $S = \varepsilon \mid S_1 \bullet S_2 \mid S_1 + S_2 \mid \mu(S_1) \mid S_1 \parallel S_2 \mid (S_1 \mid S_2) \rhd S_3 \mid (S_1)^\star \mid S_1!$ where:

- ε stands for an empty service, i.e., a service performs no operation.
- $S_1 \bullet S_2$ performs the service S_1 followed by the service S_2, i.e., $\bullet$ is an operator of sequence.
- $S_1 + S_2$ can reproduce either the behavior S_1 or S_2, i.e., $+$ is an alternative operator.
- $\mu(S_1)$ represents a composite service where the behavior S_1 may be executed multiple times, i.e., μ is an iteration operator.
- $S_1 \parallel S_2$ performs concurrently the two services S_1 and S_2 i.e., $\parallel$ is a Parallel operator.
- $(S_1 \mid S_2) \rhd S_3$ waits for the execution of one service (among the S_1 and S_2) before activating the service S_3 i.e.$\rhd$ is a discriminator operator. Note that S_1 and S_2 are executed in parallel and independently,
- $(S_1)^\star$ repressents a composite service which allows creating multiple instances of a given Web service S_1. These instances are independent of each other and run concurrently,
- $S_1!$ represents a composite service which if the Web service S_1 has started, it is disabled and, where possible, the currently running instance is halted and removed.

In this section, we give a formal definition, in term of RECATNet, of the composition operators. Let specified, as defined in *Definition* 1, each atomic Web service by $S_i = \langle NameS_i, Desc_i, Loc_i, URL_i, CS_i, RECATNetS_i\rangle$ where $RECATNetS_i = \langle Spec_i; P_i, T_i, F_i; sort_i, Cap_i, IC_i, DT_i, CT_i, TC_i, I_i, \Upsilon_i, ICT_i, K_i\rangle$. Let's define a function *initMarking*(*WS*) that is used to return the start marking i.e. initial state of the invoked Web service *WS*. The following notations are common to all the composition operators:

- *NameS* is the name of the new service,
- *Desc* is the description of the new service,
- *Loc* is the location of the new service,
- *URL* is the invocation of the new service.

5.1 Empty Service

The empty service ε is a service that performs no operation. It is used for technical and theoretical reasons.

Definition 4 The empty service is defined as $\varepsilon = \langle NameS, Desc, Loc, URL, CS, RECATNetS\rangle$ where:

- $NameS = Empty$
- $Desc =$ "*EmptyWebService*"
- $Loc = Null$ stating that there is no server for the service,
- $URL = Null$ stating that there is no *URL* for the service,
- $CS = \{Empty\}$ and
- $RECATNetS = \langle Spec; \{p\}, \phi, \phi; \phi, \phi, \phi, \phi, \phi\rangle$

In Fig. 3a, we show the graphic representation of the empty service ε.

5.2 Sequence

The sequence operator allows the construction of a service composed of two services executed one after the other. This is typically the case when a service should wait the execution result of another one before starting its execution.

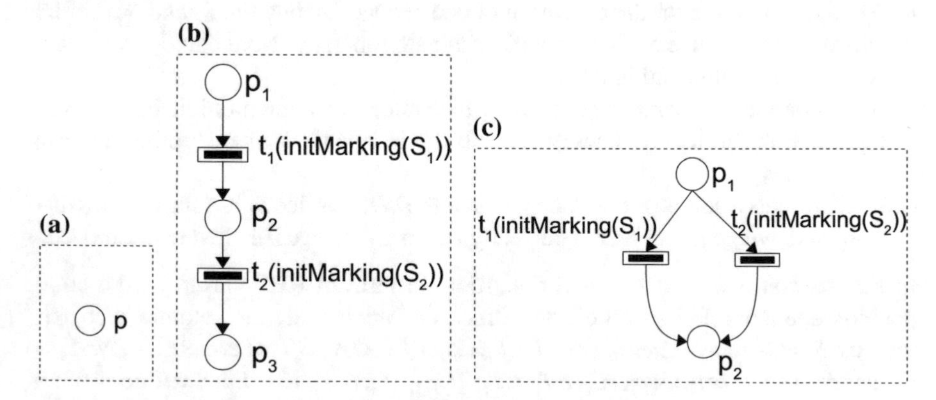

Fig. 3 Empty service (**a**), sequence service (**b**) and choice service (**c**)

Definition 5 The sequence operator $S_1 \bullet S_2$ is defined as $S = \langle NameS, Desc, Loc, URL, CS, RECATNetS \rangle$ where:

- $CS = CS_1 \cup CS_2$ and
- $RECATNetS = \langle Spec; P, T, F; ... \rangle$ where: $P = \{p_1 \cup p_2 \cup p_3\}$, $T = \{t_1 \cup t_2\}$, $F = \{(p_1, t_1), (t_1, p_2), (p_2, t_2), (t_2, p_3)\}$, $CT = \{(t_1, initMarking(S_1)), (t_2, initMarking(S_2))\}$.

Graphically, given two services S_1 and S_2, the composite service $S_1 \bullet S_2$ is represented by the RECATNet shown in Fig. 3b.

5.3 Choice

Given two services S_1 and S_2, the choice (alternative) operator allows modelling the execution of either S_1 or S_1, but not both.

Definition 6 The choice operator $S_1 + S_2$ is defined as $S = \langle NameS, Desc, Loc, URL, CS, RECATNetS \rangle$ where:

- $CS == CS_1 \cup CS_2$ and
- $RECATNetS = \langle Spec; P, T, F; ... \rangle$ where: $P = \{p_1 \cup p_2\}$, $T = \{t_1 \cup t_2\}$, $F = \{(p_1, t_1), (p_1, t_2), (t_1, p_2), (t_2, p_2)\}$, $CT = \{(t_1, initMarking(S_1)), (t_2, initMarking(S_2))\}$.

Graphically, given two services S_1 and S_2, the composite service $S_1 + S_2$ is represented by the RECATNet shown in Fig. 3c.

5.4 Iteration

The iteration operator allows a service S to be performed a certain number of times.

Definition 7 The iteration operator$\mu(S_1)$ is defined as $S = \langle NameS, Desc, Loc, URL, CS, RECATNetS \rangle$ where:

- $CS = CS_1$ and
- $RECATNetS = \langle Spec; P, T, F; ... \rangle$ where: $P = \{p_1 \cup p_2\}$, $T = \{t_1 \cup t_2\}$, $F = \{(p_1, t_1), (p_1, t_2), (t_2, p_2)\}$, $CT = \{(t_1, initMarking(S_1))\}$.

Graphically, if we consider the service S_1, the composite service $\mu(S_1)$ is represented by the RECATNet shown in Fig. 4a.

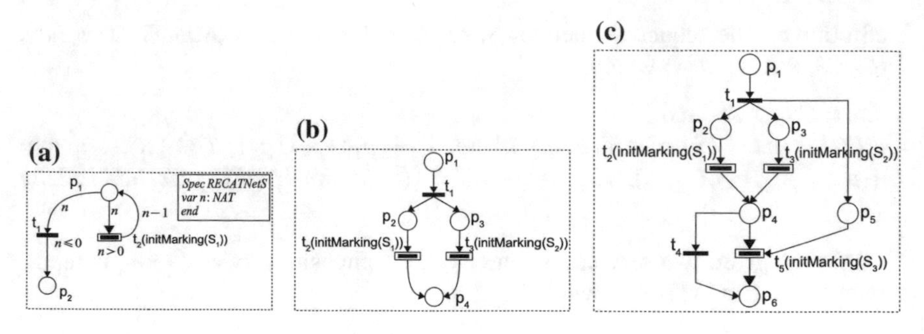

Fig. 4 Iteration (**a**), parallel (**b**) and discriminator service (**c**)

5.5 *Parallel*

Given two services S_1 and S_1, the parallel operator builds a composite service performing the two services in parallel and without interaction between them. The accomplishment of the resulting service is achieved when the two services are completed.

Definition 8 The parallel operator $S_1 \parallel S_2$ is defined as $S = \langle NameS, Desc, Loc, URL, CS, RECATNetS \rangle$ where:

- $CS = CS_1 \cup CS_2$ and
- $RECATNetS = \langle Spec; P, T, F; ... \rangle$ where: $P = \{p_1 \cup p_2 \cup p_3 \cup p_4\}$, $T = \{t_1 \cup t_2 \cup t_3\}$, $F = \{(p_1, t_1), (t_1, p_2), (t_1, p_3), (p_2, t_2), (p_3, t_3), (t_3, p_4), (t_3, p_4)\}$, $CT = \{(t_2, initMarking(S_1)), (t_3, initMarking(S_2))\}$.

Graphically, if we consider two services S_1 and S_2, the composite service $S_1 \parallel S_2$ is represented by the RECATNet shown in Fig. 4b.

5.6 *Discriminator*

Two or more equivalent services are invoked in parallel to achieve a given task but only one is required to finish before proceeding with the invocation of the next composed services of the composite service. It is presumed that these services are equivalent in terms of functionalities. The results of the first service to finish are used while the results of the remaining invoked services are ignored. At least one service of the invoked set of services must succeed for the composite service to succeed. The main goal of the discriminator operator is to increase reliability and delays of the services through the Web. For the customers, best services are those which respond in optimal time and are constantly available.

Definition 9 The discriminator operator $(S_1 \mid S_2) \triangleright S_3$ is defined as $S = \langle NameS, Desc, Loc, URL, CS, RECATNetS \rangle$ where:

- $CS = CS_1 \cup CS_2 \cup CS_3$ and
- $RECATNetS = \langle Spec; P, T, F; ...\rangle$ where: $P = \{p_1 \cup p_2 \cup p_3 \cup p_4 \cup p_5 \cup p_6\}$, $T = \{t_1 \cup t_2 \cup t_3 \cup t_4 \cup t_5\}$, $F = \{(p_1, t_1), (t_1, p_2), (t_1, p_3), (t_1, p_5), (t_2, p_4), (t_3, p_4), (p_4, t_4), (p_4, t_5, (p_5, t_5), (t_4, p_6), (t_5, p_6))\}$, $CT = \{(t_2, initMarking(S_1)), (t_3, initMarking(S_2)), (t_5, initMarking(S_3))\}$.

Graphically, if we consider three Web services S_1, S_2 and S_3, the composite service $(S_1 \mid S_2) \rhd S_3$ is represented by the RECATNet shown in Fig. 4c.

5.7 *Multiple Instance Service*

Multiple instance operator allows for a given Web service to be instantiated multiple times in a business process. The number of instances is not known during the design or run time. These instances are run concurrently but, whilst they are running, new ones can be created.

Definition 10 Multiple instance service operator $(S_1)^\star$ is defined as $S = \langle NameS, Desc, Loc, URL, CS, RECATNetS\rangle$ where:

- $CS = CS_1$ and
- $RECATNetS = \langle Spec; P, T, F; ...\rangle$ where: $P = \{p_1 \cup p_2 \cup p_{create} \cup p_{stopcreate}\}$, $T = \{t_1 \cup t_{addIns} \cup t_{remove}\}$, $F = \{(p_1, t_1), (t_1, p_2), (p_1, t_{addIns}), (t_{addIns}, p_1), (p_{create}, t_{addIns}), (p_{create}, t_{remove}), (t_{remove}, p_{stopcreate})\}$, $CT = \{(t_1, initMarking(S_1))\}$.

Graphically, if we consider a Web service S_1, the composite service $(S_1)^\star$ is represented by the RECATNet shown in Fig. 5a.

5.8 *Cancel Service*

The cancel service operator provides the ability to stop a running instance of a Web service. For instance, the purchaser can cancel his buyonline's order at any time before it starts or during its running but not after the payment was done.

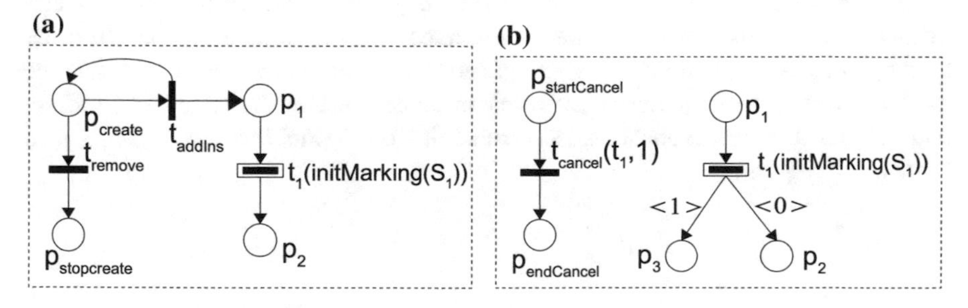

Fig. 5 Multiple instance service (**a**) and Cancel service (**b**)

Definition 11 Cancel service operator $S_1!$ is defined as $S = \langle NameS, Desc, Loc, URL, CS, RECATNetS\rangle$ where:

- $CS = CS_1$ and
- $RECATNetS = \langle Spec; P, T, F; ...\rangle$ where: $P = \{p_1 \cup p_2 \cup p_3 \cup p_{startCancel} \cup p_{endCancel}\}$, $T = \{t_1 \cup t_{cancel}\}$, $F = \{(p_1, t_1), (t_1, p_2, \langle 0\rangle), (t_1, p_3, \langle 1\rangle), (p_{startCancel}, t_{cancel}), (t_{cancel}, p_{endCancel})\}$,
  $CT = \{(t_1, initMarking(S_1))\}$, $K(t_{cancel}) = (t_1, 1)$.

Graphically, if we consider a Web service S_1, the composite service $S_1!$ is represented by the RECATNet shown in Fig. 5b.

6 A Case Study

Figure 6 shows an illustrative example of modelling a simplified *BuyOnline* service adapted from [16]. *BuyOnline* web service provides online book buying service which is composed of four atomic services: *LocateBook*, *SigIn*, *CreateAcct* and *Payement*. The composite web service may receive a list of *request*, sent by users, through the Place *StartBO*. Each *request* is represented by a token (ID, BN, SII, CAI, CCI) denotes repsectivelly *Identifier*, *BookName*, *SignInInfo*, *CreateAcctInfo*, *Credit CardInfo*. At the beginning, *BuyOnline* service starts by searching a book in web site according to the book name using service *LocateBook*. This operation is performed by firing the abstract transition *LocateBook* which, if this book can be found, returns its ISBN number. Then, the user can by this book but a valid register is required. If the user has a legal account, then finish loging using service *SignIn*; otherwise the user needs to create a new account using the service *CreateAcct*. In the last one, informations about the created account must be returned i.e. $CAO \neq \phi$. Finally, the service *payment* can finish the payment for the book according to *ISBN* number and credit card information *CCI* provided by user. Two cases are distinguished, if the credit card information are valid, the service *payment* will perform the payment by success; otherwise the service *payment* terminates by error i.e. *echecP*, and an error meaasge $CCI - not - valid$ is sent to user. Note that the user can cancel his/her online book buying service by firing the elementary transition *CancelBO*. Note that for each *request* sent by user, an instance of *BuyOnline* service is created. These instances are independent and may be executed in a distributed manner. In order to support dynamic creation instances of *BuyOnline* service, we need to update the model according to the pattern of Multiple instance operator shown in Fig. 5a. Here, and in order to perform analysis, we assume that our model is finite i.e. starts by a finite set of requests.

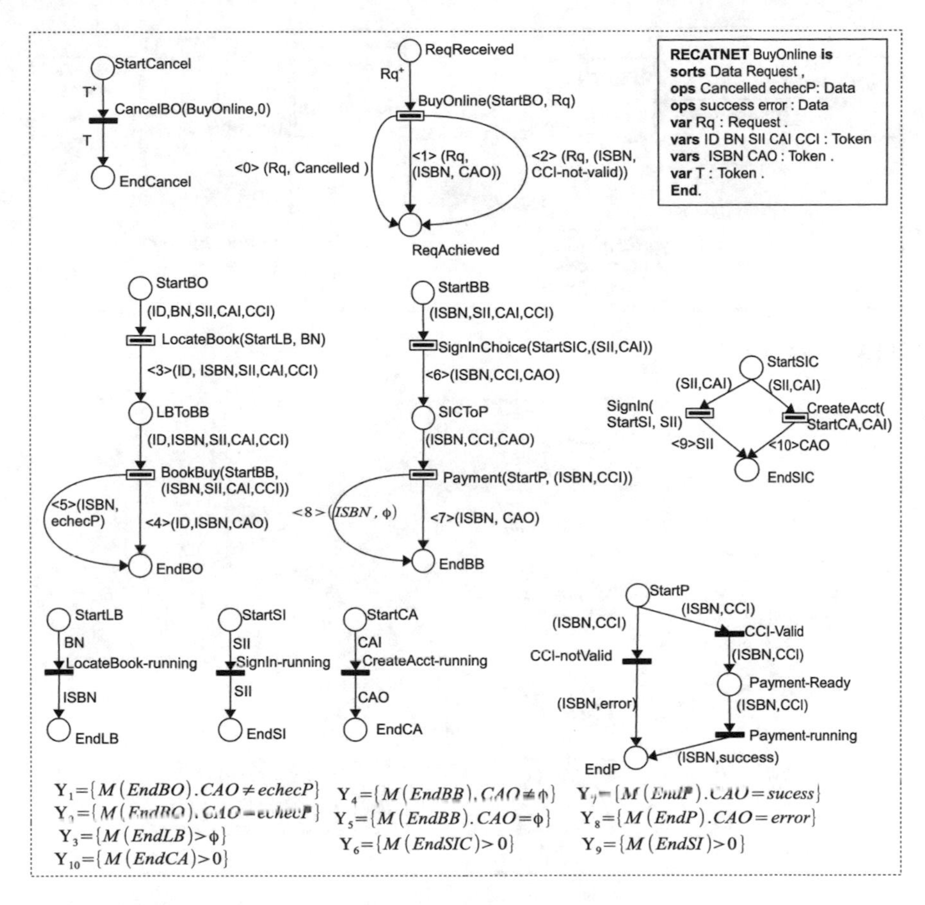

Fig. 6 *BuyOnline* book service

7 Verification of Web Services Composition

Our approach of verification can be described in Fig. 7. First, atomic services must be described in their associated RECATNets. Then, based on composition rules defined previously, generates the composite web services in terms of RECATNet. This operation may be insured by our java's tool *RECATNet-WSC* that is partially implemented. After that, from the obtained RECATNet, we generate in an automated manner its semantics in terms of rewriting logic [7] using the model-to-text (M2T) transformation tool Acceleo.[1] The rewriting logic files are used as an input of the model-checker Maude [8] to investigate several behavioral properties of Web services composition.

[1]http://www.eclipse.org/acceleo/.

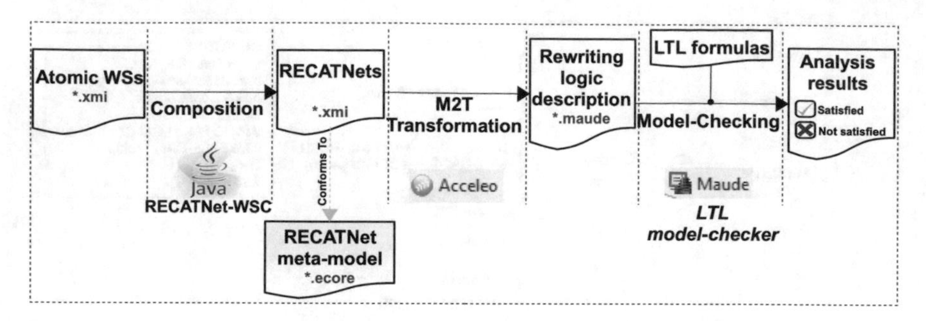

Fig. 7 Our approach

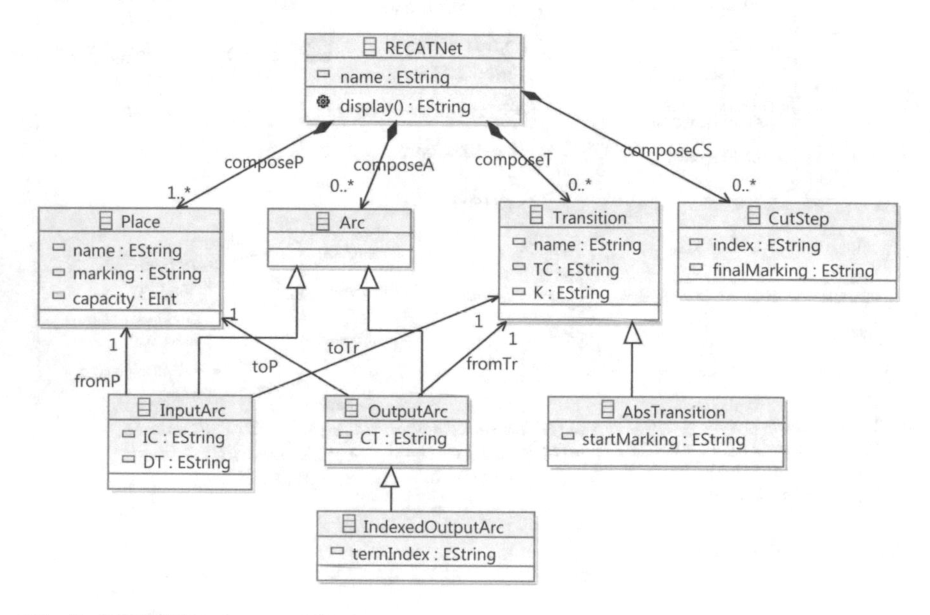

Fig. 8 RECATNets meta-model

7.1 RECATNets Meta-Modeling

In order to use M2T transformation using tool Acceleo, we need to define the meta-model of RECATNet. As shown in Fig. 8, we propose a general meta-model of our formalism using the UML class diagram model. Our proposed meta-model is composed mainly of the following classes.

- **RECATNet**: it builds the final model from a set of *Place*, *Arc*, *Transition* and *CutStep*.
- **Place**: it represents the RECATNet places. It has three attributes :*name*, *marking* and *capacity*.

- **Transition**: it represents the RECATNet transitions. It has three attributes: *name*, *TC* and *K*. One classe inherits from the super-class *Transition*: *AbsTransition* for abstract transition. This class contains one attribute *startMarking*.
- **Arc**: it represents the RECATNet arcs. It contains one attribute *inscription*. This class is a super-class of two classes. The first one is *InputArc* for arcs going from places to transitions. It contains two attributes *IC* for Input Condition and *DT* for Destroyed Token. The second is *OutputArc* for arcs going from transitions to places. It contains only the attribute *CT* for Created Token. In addition, the last class is a super-class of *IndexedOutputArc* for arcs going from abstract transition to places. It contains one attribute *index* to identify the set of indices of termination.
- **CutStep**: it represents the RECATNet cut steps. It contains two attributes *index* to identify the index of termination and *condition* to identify the condition for firing the cut step.

7.2 RECATNet Semantics in Terms of Rewriting Logic

RECATNet's semantics may be defined, easily, in terms of rewriting logic, therefore someone can use the LTL model-checker of MAUDE to investigate several behavioral properties of Web services compositions. A set of rewriting rules has been introduced in [6, 15] in order to express the semantics of RECATNet in terms of rewriting rules. In order to automate this approach, we have developed a model-to-text (M2T) transformation tool based Acceleo generator code. The transformation's rules have been inspired from rewriting rules proposed in [6]. For instance, if we consider the RECATNet of the atomic service *Payment* in Fig. 6, the generated Maude specification using M2T transformation is shown in Fig. 9. In fact, three rewriting rules are generated associated to the three elementary transitions in the RECATNet of the atomic service *Payment* in Fig. 6. For instance, the rewriting rule in line *4 rl[Payment-running]* describes the firing of the elementary transition *Payment-running*. So, this rewrite rule requires that the left-hand side is a marking where the place *Payment-Ready* is marked and yields to a marking i.e. the right-hand side, where the place *EndP* is marked.

```
1 mod PaymentS is
2 pr TYPE .
3 crl[CCI-Valid]: <StartP ; ISBN (+) CCI>=><Payment-Ready ; ISBN (+) CCI> if isValid(CCI) .
4 rl[Payment-running]: <Payment-Ready ; ISBN (+) CCI> => <EndP ; ISBN (+) success> .
5 crl[CCI-notValid]: <StartP ; ISBN (+) CCI>=><EndP ; ISBN (+) error> if not isValid(CCI) .
6 endm
```

Fig. 9 Generated Maude specification

7.3 Implementation Using the Maude Tool

An important property will be checked in Web service composition called *soundness* which concerns the correctness of internal control-flow of a composite Web service. The *soundness* of a Web services component is based on two criteria:

- *Proper termination*: This property called also *compatibility* of component Web services [17]. *Proper termination* means that starting from an initial extended marking, every possible execution path properly terminates (eventually) i.e. reaches a final extended marking. This property is expressed in LTL by the following formula: *F finalState* where the proposition *finalState* is valid in extended marking *Tr* if this latter is reduced to its root node with only terms in place *ReqAchieved*. The temporal operator *F* is denoted by $\langle\rangle$ in MAUDE notation. This formula has been proven to be *true* by MAUDE LTL-model checker in Fig. 10.
- *No dead service*: This property means that every atomic Web service must be invoked, at least, once. This requirement imposes that the Web services component should not contain Web services that can never be executed. In order to check this property, we define the proposition *isInvoked*(*ws*) which is valid in an extended marking *Tr*, if the specified Web service *ws* is invoked i.e. is running. Thus, to check that there is no dead service, we express the negation of this formula as the following LTL formula $\bigvee_{ws \in WS} G \neg isInvoked(ws)$ where *WS* is the set of atomic web services used during composition. Here, $WS = \{LocateBook, SigIn, CreateAcct, Payement\}$. If this formula is not valid, it means that the property *No dead service* is verified. The temporal operators *G* (Generally) and $\neg$ (not) are denoted, respectively, by [] and $\sim$ in MAUDE notation. In our case study, as we have a choice between the service *SigIn* and *CreateAcct*, this formula is expressed in LTL as following: $[] \sim isInvoked(LocateBook)\backslash/[] \sim (isInvoked(SignIn)\backslash/isInvoked(CreateAcct))\backslash/ [] \sim isInvoked(Payment)$. This formula has been proven to be *not valid* by MAUDE LTL-model checker in Fig. 11. The model-checker returns the expected *counterexample*.

As the two properties *Proper termination* and *No dead service* are proved to be valid, therefore, the generated composite Web service is *sound*.

```
Maude> load BuyOnline/MAIN.maude .
==========================================
reduce in RECATNET-CHECK : modelCheck(initialState, <> finalState) .
rewrites: 324827 in 14437341288ms cpu (10079ms real) (0 rewrites/second)
result Bool: true
```

Fig. 10 Checking *Proper termination* property under Maude

```
Maude> load BuyOnline/MAIN.maude .
==========================================
reduce in RECATNET-CHECK : modelCheck(initialState, []~ isInvoked(Payment) \/ (
   []~ isInvoked(LocateBook) \/ []~ (isInvoked(SignIn) \/ isInvoked(
   CreateAcct)))) .
rewrites: 183 in 8811653550ms cpu (12ms real) (0 rewrites/second)
result ModelCheckResult: counterexample(
...
{[ < ReqReceived ; ems >(*)< ReqAchieved ; 80 isbn2 CCI-not-valid empty empty
   >(*)< ListISBN ; ems >(*)< ListCAO ; ems >,nullTrans,[em,BuyOnline,[<
   SICToP ; 7(+)sii1 >,BookBuy,[< StartP ; isbn1(+)cci1 >,Payment,
   nullThread]]] [em,BuyOnline,[< SICToP ; 100(+)sii3 >,BookBuy,[< StartP ;
   isbn3(+)cci3 >,Payment,nullThread]]] [< EndBO ; 120(+)isbn4(+)echecP >,
   BuyOnline,nullThread]],'cut-2}, {[< ReqReceived ; ems >(*)< ReqAchieved ; (
   80 isbn2 CCI-not-valid empty empty)(+)(120 isbn4 CCI-not-valid empty empty)
   >(*)< ListISBN ; ems >(*)< ListCAO ; ems >,nullTrans,[em,BuyOnline,[<
   SICToP ; 7(+)sii1 >,BookBuy,[< StartP ; isbn1(+)cci1 >,Payment,
   nullThread]]] [em,BuyOnline,[< SICToP ; 100(+)sii3 >,BookBuy,[< StartP ;
   isbn3(+)cci3 >,Payment,nullThread]]]],deadlock})
```

Fig. 11 Checking *No dead service* property under Maude

8 Conclusion

In this paper, an effecient and flexible approach for Web services composition has been proposed. This approach takes fully advantage of modular, distributed execution aspects of RECATNets formalism. The formal semantic of the composition operators is expressed easily in terms of RECATNets by providing a direct transformation of each operator in terms of RECATNets. In fact, the model of RECATNets is particularly adequate for handling the most advanced flow patterns such as dynamic creation of processes and specifying exceptional behaviors in WSC at design time. Also, our method allows the verification of some properties using the LTL model-checker of the Maude system. In the future, we plan to complete this work by developing a tool capable of making automatic the mapping WSDL-descriptions into RECATNets.

References

1. Berardi, D., Calvanese, D., Giacomo, G., Lenzerini, M., & Mecella, M. (2003). Automatic composition of e-services that export their behavior. *Service-Oriented Computing—ICSOC 2003* (Vol. 2910, pp. 43–58), series Lecture Notes in Computer Science.
2. Lucchi, R., & Mazzara, M. (2007). A pi-calculus based semantics for ws-bpel. *The Journal of Logic and Algebraic Programming*, *70*(1), 96–118.

3. Hamadi, R., & Benatallah, B. (2003). A petri net-based model for web service composition. *Proceedings of the 14th Australasian Database Conference* (Vol. 17, pp. 191–200), series ADC '03.
4. Zhang, Z.-L., Hong, F., & Xiao, H.-J. (2008). A colored petri net-based model for web service composition. *Journal of Shanghai University (English Edition)*, *12*(4), 323–329.
5. Haddad, S., & Poitrenaud, D. (2007). Recursive petri nets: Theory and application to discrete event systems. *Acta Informatica*, *44*(7), 463–508.
6. Barkaoui, K., & Hicheur, A. (2008). Towards analysis of flexible and collaborative workflow using recursive ecatnets. In: A. Hofstede, B. Benatallah & H.-Y. Paik (Eds.), *Business Process Management Workshops* (vol. 4928, pp. 232–244), series Lecture Notes in Computer Science.
7. Bruni, R., & Meseguer, J. (2006). Semantic foundations for generalized rewrite theories. *Theoretical Computer Science*, *360*(1), 386–414.
8. Clavel, M. et al. (2007). Maude manual (version 2.3). http://maude.cs.uiuc.edu.
9. Srivastava, B., & Koehler, J. (2003). Web service composition—current solutions and open problems. *In: ICAPS 2003 Workshop on Planning for Web Services* (pp. 28–35).
10. Casati, F., Ilnicki, S., Jin, L.-J. & Shan, M.-C. (2000). *An Open, Flexible, and Configurable System for Service Composition* (pp. 125–132).
11. Adam, N. R., Atluri, V., & Huang, W.-K. (1998). Modeling and analysis of workflows using petri nets. *Journal of Intelligent Information Systems*, *10*(2), 131–158.
12. Jensen, K. (1990). Coloured petri nets: A high level language for system design and analysis. Technical Report.
13. Chemaa, S., Elmansouri, R., & Chaoui, A. (2013). Web services modeling and composition approach using object-oriented petri nets. *CoRR*, abs/1304.2080.
14. Russell, N., ter Hofstede, A., van der Aalst, W., & Mulyar, N. (2006). Workflow control-flow patterns: A revised view, BPM Center, Technical Report BPM-06-22.
15. Barkaoui, K., Boucheneb, H., & Hicheur, A. (2009). *Modelling and Analysis of Time-constrained Flexible Workflows with Time Recursive Ecatnets* (vol. 5387, pp. 19–36), series Lecture Notes in Computer Science. Berlin, Heidelberg: Springer.
16. Ding, Z., Wang, J., & Jiang, C. (2008). An approach for synthesis petri nets for modeling and verifying composite web service. *Journal of Information Science and Engineering* 1309–1328.
17. Li, X., Fan, Y., Sheng, Q., Maamar, Z., & Zhu, H. (2011). A petri net approach to analyzing behavioral compatibility and similarity of web services. *IEEE Transactions on Systems, Man and Cybernetics, Part A: Systems and Humans*, *41*(3), 510–521.

PMRF: Parameterized Matching-Ranking Framework

Fatma Ezzahra Gmati, Nadia Yacoubi-Ayadi, Afef Bahri, Salem Chakhar and Alessio Ishizaka

Abstract The PMRF (Parameterized Matching-Ranking Framework) is a highly configurable framework supporting a parameterized matching and ranking of Web services. This paper first introduces the matching and ranking algorithms supported by the PMRF. Next, it presents the architecture of the developed system and discusses some implementation issues. Then, it provides the results of performance evaluation of the PMRF. It also compares PMRF to two exiting frameworks, namely iSeM-logic-based and SPARQLent. The different matching and ranking algorithms have been evaluated using the OWLS-TC4 datasets. The evaluation has been conducted employing the SME2 (Semantic Matchmaker Evaluation Environment) tool. The results show that the algorithms behave globally well in comparison to iSeM-logic-based and SPARQLent.

F.E. Gmati · N. Yacoubi-Ayadi
RIADI Research Laboratory, National School of Computer Sciences,
University of Manouba, 2010, Manouba, Tunisia
e-mail: fatma.ezzahra.gmati@gmail.com

N. Yacoubi-Ayadi
e-mail: nadia.yacoubi.ayadi@gmail.com

A. Bahri
MIRACL Laboratory, Higher School of Computing and Multimedia,
Technopole Sfax, 3021 Sfax, Tunisia
e-mail: afef.bahri@gmail.com

S. Chakhar (✉) · A. Ishizaka
Portsmouth Business School and Centre for Operational Research and Logistics,
University of Portsmouth, Portland Street, Portsmouth PO1 3AH, UK
e-mail: salem.chakhar@port.ac.uk

A. Ishizaka
e-mail: alessio.ishizaka@port.ac.uk

R. Lee (ed.), *Software Engineering Research, Management and Applications*, Studies in Computational Intelligence 654,
DOI 10.1007/978-3-319-33903-0_13

1 Introduction

The matchmaking is a crucial operation in Web service composition. The objective of the matchmaking is to discover and select the most appropriate (i.e., that responds better to the user request) Web service among the different available candidates. Several matchmaking frameworks are now available in the literature, e.g., [1, 14, 16, 17, 19, 20, 23, 24, 26–28]. However, most of these frameworks present at least one of the following shortcomings:

1. use of strict syntactic matching, which generally leads to low recall and low precision of the retrieved services;
2. use of capability-based matchmaking, which is proven [6] to be inadequate in practice;
3. lack of customization and configurability support for both the user and the provider;
4. lack of accurate ranking of matching Web services, especially within semantic-based matching.

Several conceptual and algorithmic solutions to jointly deal with the previous shortcomings are under investigation in an ongoing research project. The first results are given in [8]. The objective of this paper is to present the developed prototype, PMRF (Parameterized Matching-Ranking Framework), supporting the different proposed matching and ranking algorithms. The paper first introduces the matching and ranking algorithms supported by the PMRF. Then, it presents the architecture of the developed system and discusses some implementation issues. Finally, it provides the results of performance evaluation of the PMRF and also compares it to two well-known matchmakers, namely iSeM-logic-based [11] and SPARQLent [21, 22].

To evaluate the performance of PMRF, we used seven different configurations with different versions of matching and ranking algorithms. All the algorithms have been evaluated using the OWLS-TC4 datasets. The evaluation has been conducted employing the SME2 (Semantic Matchmaker Evaluation Environment) tool [12]. The results show that the algorithms behave globally well in comparison to iSeM-logic-based and SPARQLent.

The rest of the paper is organized as follows. Section 2 reviews the matching and ranking algorithms. Section 3 presents the architecture of the PMRF. Section 4 studies the performance of the PMRF. Section 5 compares the PMRF to other similar frameworks. Section 6 comments on the users/providers acceptability. Section 7 discusses some related work. Section 8 concludes the paper.

2 Matching and Ranking Algorithms

In this section, we briefly review the matching and ranking algorithms supported by the PMRF.

2.1 Matching Algorithms

The PMRF supports three matching algorithms: trivial, partially parameterized and fully parameterized. These algorithms support different levels of customization. The trivial matching algorithm supports no customization. The partially parameterized matching algorithm allows the user to specify the set of attributes to be used in the matching. Within the fully parameterized matching algorithm, three customizations are taken into account: (i) A first customization consists in allowing the user to specify the list of attributes to consider; (ii) A second customization consists in allowing the user to specify the order in which the attributes are considered; and (iii) A third customization is to allow the user to specify a desired similarity measure for each attribute. In the rest of this section, we present the third algorithm.

In order to support all the above-cited customizations, we used the concept of Criteria Table, introduced by [6], that serves as a parameter to the matching process. A Criteria Table, C, is a relation consisting of two attributes, $C.A$ and $C.M$. $C.A$ describes the service attribute to be compared, and $C.M$ gives the *least preferred similarity measure* for that attribute. Let $C.A_i$ and $C.M_i$ denote the service attribute value and the desired measure in the ith tuple of the relation. $C.N$ denotes the total number of tuples in C.

Let S^R be the service that is requested, and S^A be the service that is advertised. Let C be a criteria table. A sufficient match exists between S^R and S^A if for *every* attribute in $C.A$ there exists an identical attribute of S^R and S^A and the values of the attributes satisfy the desired similarity measure as specified in $C.M$. Formally,

$$\begin{aligned}&\forall_i \exists_{j,k}(C.A_i = S^R.A_j = S^A.A_k) \wedge \mu(S^R.A_j, S^A.A_k) \geq C.M_i\\&\Rightarrow \mathrm{SuffMatch}(S^R, S^A) \quad 1 \leq i \leq C.N.\end{aligned} \tag{1}$$

According to this definition, only the attributes specified by the user in the Criteria Table are considered during the matching process.

The fully parameterized matching algorithm is formalized in Algorithm 1. This algorithm follows directly from Sentence (1). Algorithm 1 proceeds as follows: (i) Loops over the Criteria Table and for each attribute it identifies the corresponding attribute in the requested service S^R and the potentially advisable service under consideration S^A. The corresponding attributes are appended into two different lists rAttrSet (for requested Web service S^A) and aAttrSet (for advisable Web service S^A). This operation is implemented by sentences 1–10 in Algorithm 1; and (ii) Loops over the Criteria Table and for each attribute it computes the similarity degree between the corresponding attributes in rAttrSet and aAttrSet. This operation is implemented by sentences 11–14 in Algorithm 1. The output of Algorithm 1 is either success (if for every attribute in the Criteria Table C there are similar attribute in the advertised service S^A with a sufficient similarity degree) or fail (if the similarity for at least one attribute in the Criteria Table C fails).

Let us now focus on the complexity of Algorithm 1. Generally, we have $S^A.N \gg S^R.N$, hence the complexity of the first outer *while* loop is $O(C.N \times S^A.N)$. Then, the worst case complexity of Algorithm1 is $O(C.N \times S^A.N) + \alpha$ where α is the complexity of computing μ. The value of α depends on the approach used to infer $\mu(\cdot,\cdot)$. As underlined in [6], inferring $\mu(\cdot,\cdot)$ by ontological parse of pieces of information into facts and then utilizing commercial rule-based engines, which use the fast Rete [7] pattern-matching algorithm leads to $\alpha = O(|R||F||P|)$ where $|R|$ is the number of rules, $|F|$ is the number of facts, and $|P|$ is the average number of patterns in each rule. In this case, the worst case complexity of Algorithm 1 is $O(C.N \times S^A.N) + O(|R||F||P|)$. Furthermore, we observe, as in [6], that the process of computing $\mu(\cdot,\cdot)$ is the most 'expensive' step of Algorithm 1. Hence, we obtain: $O(C.N \times S^A.N) + O(|R||F||P|) \asymp O(|R||F||P|)$.

Algorithm 1: Fully Parameterized Matching

```
Input  : S^R, // Requested service.
         S^A, // Advertised service.
         C, // Criteria Table.
Output: Boolean, // fail/success.
1  while (i ≤ C.N) do
2      while (j ≤ S^R.N) do
3          if (S^R.A_j = C.A_i) then
4              Append S^R.A_j to rAttrSet;
5          j ⟵ j + 1;
6      while (k ≤ S^A.N) do
7          if (S^A.A_k = C.A_i) then
8              Append S^A.A_k to aAttrSet;
9          k ⟵ k + 1;
10     i ⟵ i + 1;
11 while (t ≤ C.N) do
12     if (μ(rAttrSet[t], aAttrSet[t]) ≺ C.M_t) then
13         return fail;
14     t ⟵ t + 1;
15 return success;
```

Different versions and extensions of this algorithm are available in [4, 5, 8]. We remark that Algorithm 1 permits to compute the similarly between a requested Web service S^R and an advertised Web service S^A. In practice, however, matching process should consider all the Web services available in the registry. An extended version of Algorithm 1 that takes into account this fact is given in [8].

2.2 *Ranking Algorithms*

The PMRF supports three ranking algorithms: score-based, rule-based and tree-based. The first algorithm relies on the scores only. The second algorithm defines and uses a series of rules to rank Web services. It permits to solve the ties problem encountered by the score-based ranking algorithm. The tree-based algorithm is based on the use of a tree data structure. It permits to solve the problem of ties of the first algorithm. In addition, it is computationally better than the rule-based ranking algorithm. In the present paper, we present the score-based ranking algorithm. We note that the rule-based ranking algorithm is available in [8] while the tree-based algorithm is given in [9].

The score-based ranking approach is implemented by Algorithm 2. The main input of this algorithm is a list mServices of matching Web services. The function ComputeNormalizedScores in Algorithm 2 permits to calculate the scores of Web services. It implements the idea we proposed in [8]. The score-based ranking algorithm uses then a *merge sort procedure* (implemented by lines 3–11 in Algorithm 2 to rank the Web services based on their normalized scores.

The list mServices used as input to Algorithm 2 has the following generic definition:

$$(S_i^A, \mu(S_i^A.A_1, S^R.A_1), \ldots, \mu(S_i^A.A_N, S^R.A_N)),$$

where: S_i^A is an advertised Web service, S^R is the requested Web service, N the total number of attributes and $\mu(S_i^A.A_j, S^R.A_j)$ $(j = 1, \ldots, N)$ is the similarity measure between the requested Web service and the advertised Web service on the jth attribute A_j.

The list mServices will be first updated by ComputeNormalizedScores and it will have the following new generic definition:

$$(S_i^A, \mu(S_i^A.A_1, S^R.A_1), \ldots, \mu(S_i^A.A_N, S^R.A_N), \rho'(S_i^A)),$$

where: S_i^A, S^R, N and $\mu(S_i^A.A_j, S^R.A_j)$ $(j = 1, \ldots, N)$ are as described above; and $\rho'(S_i^A)$ is the normalized score of advertised Web service S_i^A.

Algorithm 2: Score-Based Ranking

```
   Input  : mServices, // List of matching Web services.
            N, // Number of attributes.
   Output: mServices, // Ranked list of Web services.
1  mServices ← ComputeNormalizedScores(mServices,N) ;
2  r ← length(mServices);
3  while (i ≤ r) do
4      Let row_i be the ith row in mServices ;
5      while (j ≤ r) do
6          Let row_j be the jth row in mServices ;
7          if (mServices[i, N + 2] > mServices[j, N + 2])) then
8              tmp ⟵ row_j;
9              row_j ⟵ row_i;
10             row_i ⟵ tmp;
11             update mServices ;

12 return mServices ;
```

Two versions can be distinguished for the definition of the list mServices at the input level, along with the way the similarity degrees are computed. The first version of mServices is as follows:

$$(S_i^A, \mu_{\max}(S_i^A.A_1, S^R.A_1), \ldots, \mu_{\max}(S_i^A.A_N, S^R.A_N)),$$

where: S_i^A, S^R and N are as defined above; and $\mu_{\max}(S_i^A.A_j, S^R.A_j)$ $(j = 1, \ldots, N)$ is the similarity measure between the requested Web service and the advertised Web service on the jth attribute A_j. In this case, the similarity measure is computed by selecting the edge with the **maximum weight** in the matching graph.

The second version of mServices is as follows:

$$(S_i^A, \mu_{\min}(S_i^A.A_1, S^R.A_1), \ldots, \mu_{\min}(S_i^A.A_N, S^R.A_N)),$$

where S_i^A, S^R and N are as defined above; and $\mu_{\min}(S_i^A.A_j, S^R.A_j)$ $(j = 1, \ldots, N)$ is the similarity measure between the requested Web service and the advertised Web service on the jth attribute A_j. In this case, the similarity measure is computed by selecting the edge with the **minimum weight** in the matching graph.

To obtain the final rank, we need to use these two versions separately and then combine the obtained rankings. However, a problem of ties may occur since several Web services may have the same scores with both versions. This will deteriorate the precision of the ranking algorithm. The tree-based ranking algorithm permits to completely solve the ties problem.

The function ComputeNormalizedScores in Algorithm 2 has a complexity of $O(r(2 + N^2))$ where r is the number of Web services and N is the number of attributes. The length in line 2 is assumed to be a built-in function and its complexity is not considered here. The sentences in lines 3–11 in Algorithm 2 implement a merge sort procedure, which at best has a time complexity of $O(r \log r)$ and in worst case, it makes $O(r^2)$. Hence, the overall complexity of Algorithm 2 in best case is $O(r(2 + N^2)) + O(r \log r)$ and in worst case is $O(r(2 + N^2)) + O(r^2)$.

3 System Architecture and Implementation

In this section, we first present the conceptual and functional architectures of the PMRF. Then, we discuss some implementation issues.

3.1 System Design and Conceptual Architecture

Figure 1 provides the conceptual architecture of the PMRF. The inputs of the system are: the Criteria Table/List, the published Web services repository, the user request and its corresponding Ontologies. The other parameters (namely, the similarity degrees weights and the order functions; see [8]) are computed by the PMRF. The output of the PMRF is a ranked list of Web services.

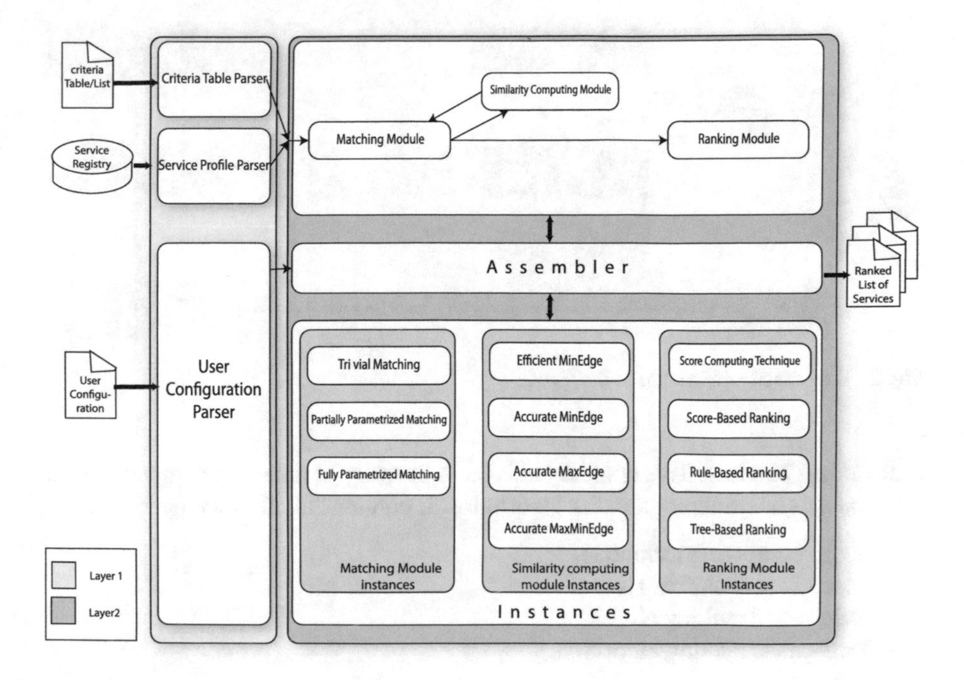

Fig. 1 Conceptual architecture of the PMRF

The PMRF is composed of two layers. The role of the first layer is to parse the input data and parameters and then transfer it to the second layer, which represents the matching and ranking engine. The Matching Module filters Web service offers that match with the Criteria Table/List. The result is then passed to the Ranking Module. This module produces a ranked list of Web services. The assembler guarantees a coherent interaction between the different modules in the second layer.

The three main components of the second layer of the PMRF are:

- **Matching Module**: This component contains the different matching algorithms:
  - Trivial matching algorithm,
  - Partially parameterized matching algorithm,
  - Fully parameterized matching algorithm.
- **Similarity Computing Module**: This component supports the different similarity measure computing approaches:
  - Efficient similarity with MinEdge,
  - Accurate similarity with MinEdg,
  - Accurate similarity with MaxEdge,
  - Accurate similarity with MaxMinEdge.

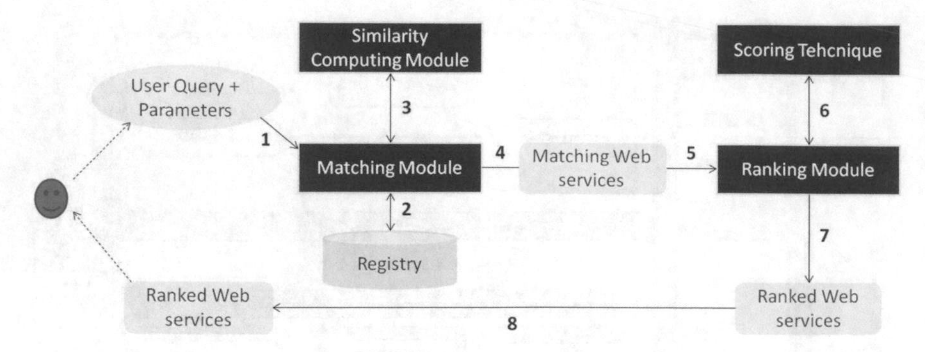

Fig. 2 Functional architecture of the PMRF

- **Ranking Module**: This component is the repository of the score computing technique and the different ranking algorithms. It contains the following elements:
  - Score computing technique,
  - Score-based ranking algorithm,
  - Rule-based ranking algorithm,
  - Tree-based ranking algorithm.

3.2 Functional Architecture

The functional architecture of the PMRF is given in Fig. 2. It shows graphically the different steps from receiving the user query (including the specifications of the requested Web service and the different parameters) until the delivery of the final results (ranked list of Web services matching the query) to the user. We can distinguish the following main operations:

- The PMRF receives (1) the user query including the specifications of the desired Web service and the required parameters;
- The Matching Module scans (2) the Registry in order to identify the Web services matching the user query;
- During the matching process, the Matching Module uses (3) the Similarity Computing Module to calculate the similarity degrees;
- The Matching Module delivers (4) the Web services matching the user query;
- The Ranking Module receives (5) the matching Web services and processes them for ranking;
- During the ranking operation, the Ranking Module uses (6) the Scoring Technique to compute the scores of the Web services;
- The Ranking Module delivers (7) a ranked list of Web services;
- The PMRF delivers (8) the ranked list of Web services to the user.

3.3 Implementation

To develop the PMRF, we have used the following tools:

- Eclipse IDE (http://eclipse.org/ide/) as the developing platform;
- OWLS-API (http://on.cs.unibas.ch/owls-api/) to parse the OWLS service descriptions;
- OWL-API (http://owlapi.sourceforge.net/) along with the Pellet reasoner (http://clarkparsia.com/pellet/) to perform the inference for computing the similarity degrees.

The inference is one of the main issues encountered during the developing of the PMRF. We perform the following procedure in order to minimize resources consumption, especially memory:

1. A local Ontology is created at the start of the matchmaking process. The incremental classifier class, taken from the Pellet reasoner library, is associated to this Ontology.
2. The service parser based on the OWLs-API retrieves the Uniform Resource Identifier (URI) of the attributes values of each service. The concepts related to these URIs are added incrementally to the local Ontology and the classifier is updated accordingly.
3. In order to infer the semantic relations between concepts, the similarity measure module uses the knowledge base constructed by the incremental classifier.

Figure 3 provides an extract from the class Matchmaker. In this figure, we can see the input and output functions. The latter contains the call for the matching and ranking operations.

4 Performance Evaluation

In this section, we provide the performance evaluation results.

4.1 Evaluation Framework

To evaluate the performance of the PMRF, we used the SME2 [12], which is an open source tool for testing different semantic matchmakers in a consistent way. The SME2 uses OWLS-TC collections to provide the matchmakers with Web service descriptions, and to compare their answers to the relevance sets of the various queries.

The SME2 provides several metrics to evaluate the performance and effectiveness of a Web service matchmaker. The metrics that have been considered in this paper are: precision and recall, average precision, query response time and memory consumption. The definition of these metrics are given in [12, 13].

```
27 public class Matchmaker implements IMatchmakerPlugin {
28
29         PelletReasoner reasoner=new PelletReasoner();
30         ServiceTuple query;
31         ArrayList<ServiceTuple> offers=new ArrayList<ServiceTuple>();
32
33
34⊖        public Matchmaker()
35         {}
36
37⊖        @Override
38         public void input(URL arg0) {
39             try {
40                     ServiceTuple service=new ServiceTuple(arg0,reasoner);
41                     offers.add(service);
42                     System.out.println("helloooo");
43             } catch (Exception e) {
44                     e.printStackTrace();
45             }
46         }
47⊖        @Override
48         public Hashtable<URL, Vector<URL>> query(URL arg0)
49         {
50         Hashtable<URL,Vector<URL>> finalOutput=new Hashtable<URL,Vector<URL>>();
51         try
52         {
53             query=new ServiceTuple(arg0,reasoner);
54             /*
55              * ***********************************************************************
56              * We first perform the matching
57              * ***********************************************************************
58              */
59             Group initialGroup=new Group();
60             for(ServiceTuple serviceAd:offers)
61             {
62                 match(query,serviceAd,reasoner);
63                 initialGroup.addAService(serviceAd);
64             }
65             /*
66              * ***********************************************************************
67              * We secondly perform the ranking
68              * ***********************************************************************
69              */
70             Node<Group>  root= new Node<Group>();
71             root.setData(initialGroup);
```

Fig. 3 Extract from the class matchmaker

A series of experimentations have been conducted on a Dell Inspiron 15 3735 Laptop with an Intel Core I5 processor (1.6 GHz) and 2 GB of memory. The test collection used is OWLS-TC4, which consists of 1083 Web service offers described in OWL-S 1.1 and 42 queries.

4.2 Performance Evaluation Analysis

In order to study the performance of each instance of the modules supported by the PMRF and describe the difference between them, we implemented seven plugins to be used with the SME2 tool. Each of these plugins represents a different combination of the matching, similarity computing and ranking algorithms. The characteristics of these plugins are summarized in Table 1.

Table 1 Description of the evaluated configurations

Configuration	Similarity measure	Matching	Ranking
1	Accurate MinEdge	Trivial	Trivial
2	Efficient MinEdge	Trivial	Trivial
3	Accurate MaxEdge	Trivial	Trivial
4	Accurate MinEdge	Fully parameterized	Trivial
5	Accurate MaxMinEdge	Trivial	RankMinMax
6	Accurate MinEdge	Trivial	Rule based
7	Efficient MinEdge	Trivial	Rule based

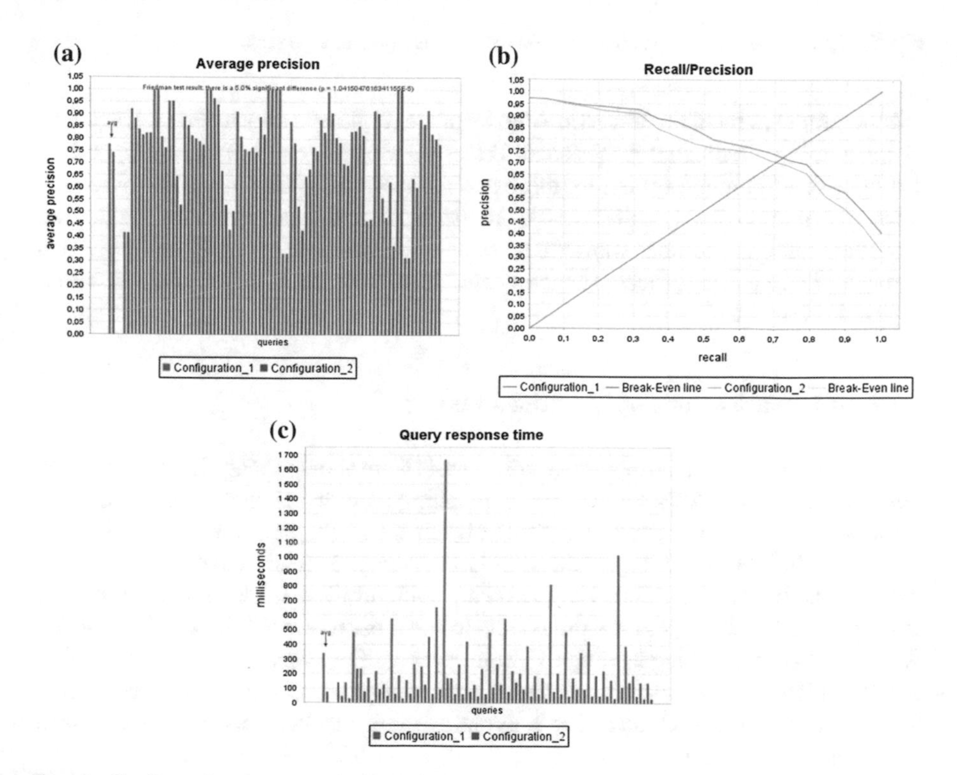

Fig. 4 Configuration 1 versus configuration 2: **a** average precision, **b** recall/precision, **c** query response time

4.2.1 Comparison of Configurations 1 and 2

The evaluation of configurations 1 and 2 yields to the results shown in Fig. 4. The difference between the two configurations is the similarity measure module instance. Indeed, the first configuration employs the **Accurate MinEdge** instance while the

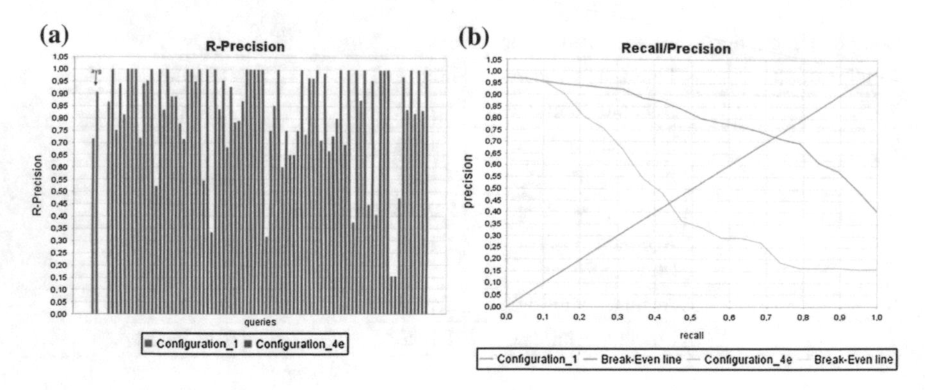

Fig. 5 Configuration 1 versus configuration 4: **a** average precision, **b** recall/precision

second employs the **Efficient MinEdge** instance. Figure 4a shows the Average Precision and Fig. 4b illustrates the Recall/Precision plot. We can see that configuration 1 outperforms configuration 2 for these two metrics, this is due to the use of logical inference, that obviously enhances the precision of the first configuration. In Fig. 4c, however, configuration 2 is shown to be remarkably faster than configuration 1. This is due to the inference process (which is used in configuration 1) that consumes considerable resources.

4.2.2 Comparison of Configurations 1 and 4

The results of comparison of configuration 1 and 4 are shown in Fig. 5. The difference between these two configurations is the matching module instance. The first configuration is based on the trivial matching algorithm while the second uses the fully parameterized matching. Figure 5a shows the Average Precision metric results. It is easy to see that configuration 4 outperforms configuration 1. This is due to the fact that the Criteria Table restricts the results to the most relevant Web services, which will have the best ranking leading to a high Average Precision value. Figure 5b illustrates the Recall/Precision plot. It shows that configuration 4 has a low recall rate. The overly restrictive Criteria Table explains these results, since it fails to return some relevant services.

4.2.3 Comparison of Configurations 5 and 6

Figure 6 show the evaluation results of configurations 5 and 6. The difference between these two configurations is the ranking module instance. The first uses the tree-based ranking algorithm while the second employs the rule-based ranking algorithm. Figure 6a shows that configuration 5 has a slightly better Average Precision than configuration 6. Figure 6b shows that configuration 6 is obviously faster than configuration 5.

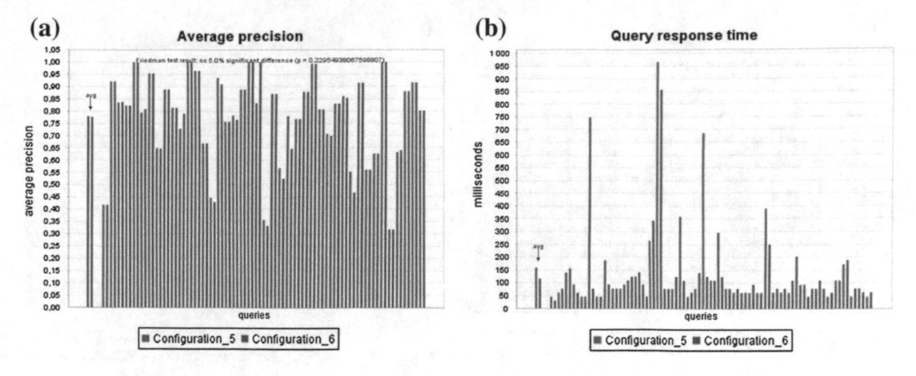

Fig. 6 Configuration 5 versus configuration 6: **a** average precision, **b** query response time

5 Comparative Study

We compared the results of the PMRF matchmaker with SPARQLent [21, 22] and iSeM [11] frameworks. Configuration 7 was chosen to perform this comparison. The SPARQLent is a logic-based matchmaker based on the OWL-DL reasoner Pellet to provide exact and relaxed Web services matchmaking. The iSeM is an hybrid matchmaker offering different filter matchings: logic-based, approximate reasoning based on logical concept abduction for matching Inputs and Outputs. We considered only the I-O logic-based in this comparative study. We note that SPARQLent and iSeM consider preconditions and effects of Web services, which are not considered in our work.

5.1 Average Precision

The Average Precision is shown in Fig. 7a. This figure shows that PMRF has a more accurate Average Precision than iSeM logic-based and SPARQLent. It is possible to conclude that PMRF has better ranking precision than the two other approaches. In addition, the ranking generated is more fine-grained than SPARQLent and iSeM. This is due to the score-based ranking that gives a more coarse evaluation than a degree aggregation. Indeed, SPARQLent and iSeM approaches adopt a subsumption-based ranking strategy as described in [18], which gives equal weights to all similarity degrees.

5.2 Recall/Precision

Figure 7b presents the Recall/Precision of PMRF, iSeM logic-based and SPARQLent. This figure shows that PMRF recall is significantly better than both iSeM

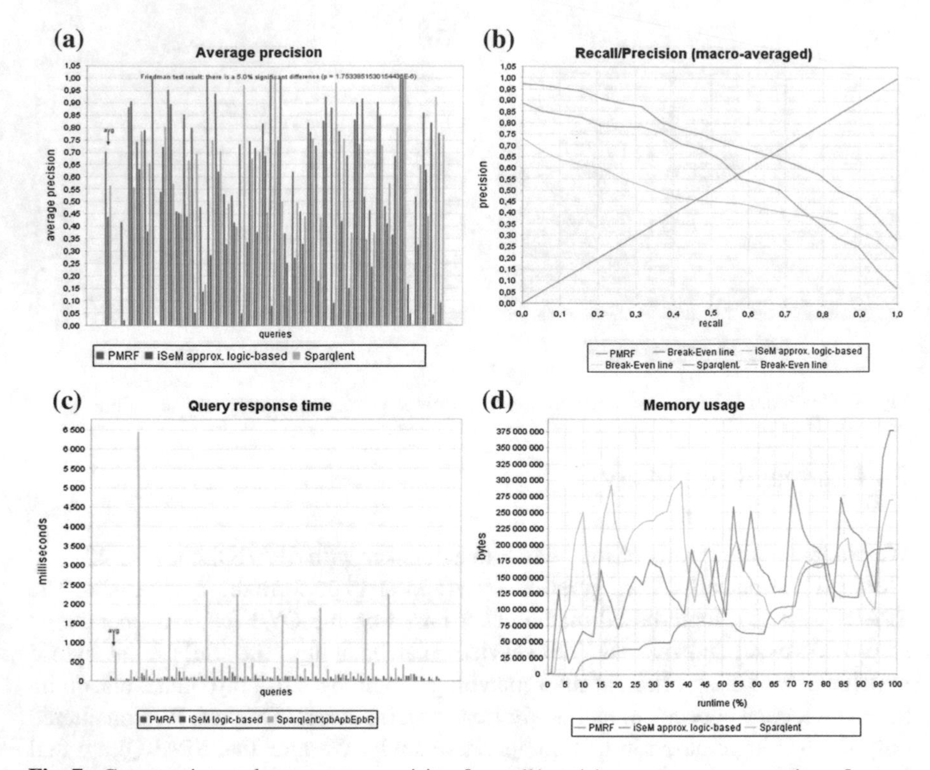

Fig. 7 Comparative study: **a** average precision, **b** recall/precision, **c** query response time, **d** memory usage

logic-based and SPARQLent. This means that our approach is able to reduce the amount of false positives (see [2] for a discussion on the false positives problem).

5.3 *Query Response Time*

Figure 7c compares the Query Response Time of the PMRF, logic-based iSeM and SPARLent. The first column (Avg) gives the average response time for the three matchmakers. The experimental results show that the PMRF is faster than SPARQLent (760 ms for SPARQLent vs. 128 ms for PMRF) and slightly less faster than logic-based iSeM (65 ms for iSeM). We note that SPARQLent has especially high query response time if the query include preconditions/effects. The SPARQLent is also based on an OWL DL reasoner, which is an expensive processing. PMRF and iSeM have close query response time because both consider direct parent/child relations in a subsumption graph, which reduces significantly the query processing. The PMRF highest query response time limit is 248 ms.

5.4 Memory Usage

Figure 7d shows the Memory Usage for PMRF, iSeM logic-based and SPARQLent. It is easy to see that PMRF consumes less memory than iSeM logic-based and SPARQLent. This can be explained by the fact that the PMRF does not require a reasoner (in the case of Configuration 7) neither a SPARQL queries in order to compute similarities between concepts.

6 Discussion

An important characteristic of the proposed framework is its configurability by allowing the user to specify a set of parameters and apply different algorithms supporting different levels of customization. This, however, leads to the problem of users/providers acceptability and their ability to specify the required parameters, especially the Criteria Table. Indeed, the specification of these parameters may require an important cognitive effort from the user/providers. A possible solution to reduce this effort is to use a predefined Criteria Table. This solution can be further enhanced by including in the framework some appropriate Artificial Intelligence techniques to learn from the previous choices of the user.

Another possible solution to reduce the cognitive effort consists in exploiting the context of the user queries. First, the description of elementary services can be textually analysed and, based on the query domain, the system uses either the efficient or the accurate configurations. Second, a global time limit to the composition process can be used to orient the system towards the use of the accurate version or efficient version of the similarity measure computing algorithm. Third, the context of the query in the workflow can be used to determine the level of customization needed and also in the generation of a suitable Criteria Table or Attributes List.

A more advanced solution consists in combining all the solutions cited above.

7 Related Work

Several existing frameworks have influenced this research project, especially the proposals of [2, 4–6, 11, 18]. Table 2 provides the characteristics of some existing frameworks. Ludwig [16] proposes two matchmaking approaches: one that is based on a genetic algorithm, and the other is based on a memetic algorithm to match consumers with services based on Quality of Service (QoS) attributes. Wang et al. [27] propose the use of utility function to evaluate each component service based on the definition given in [28] and then map the multi-dimensional QoS composite Web service to the multi-dimensional multi-choice knapsack. Finally, they use an heuristic algorithm for solving the problem.

Table 2 Comparison of matchmaking frameworks

Matchmaker	Matching type	Attributes	Customization	Ranking	Description language
Jini [1]	Syntactic	Capability	No	No	No
Konark [14]	Syntactic	Capability	No	No	XML
Salutation [17]	Logic-based	Capability	No	Yes	OWL-S
MatchMaker [25]	Syntactic	Capability	No	No	DAMS/UDDI
RACER [15]	Syntactic	Capability	No	No	DAML-S
PSMF [6]	Logic-based	Capability	Yes	No	DAML-S/WSDL/UDDI
SPARQLent [22]	Logic-based	Capability	No	Yes	OWL-S
iSeM-logic-based [11]	Logic-based	Capability	No	Yes	OWL-S/SAWSDL
QoSeBroker [5]	Logic-based	Capability/QoS/property	Yes	No	OWL-S
PMRF	Logic-based	Capability/property	Yes	Yes	OWL-S

Some proposals including [3, 10] propose to use semantics to enhance the matchmaking process but most of them still consider capability attributes only. The proposal of [4, 5] lack effective implementation of the proposed matchmaking framework. Indeed, the authors discuss very generally and very briefly the technical issues. In addition, the authors do not precise how the similarity degree is computed and how the different matching Web services are ranked. Finally, there is a lack of effective evaluation and performance analysis of matching algorithms.

Although that these proposals are based on semantics, they fail to take into account jointly the shortcomings of Web services matchmaking enumerated in the introduction. Indeed, the proposal of [2, 11, 22] do not support any customization while those of [4–6] do not propose solutions for ranking Web services.

8 Conclusion

In this paper, we presented a highly customizable framework, called PMRF, for matching and ranking Web services. We briefly reviewed the matching and ranking algorithms supported by the PMRF, provided its conceptual and functional architecture and discussed some implementation issues. We also presented the results of the performance evaluation of the PMRF using the OWLS-TC4 datasets. The evaluation has been conducted using the SME2 tool [12]. We finally compared PMRF to two exiting frameworks, namely iSeM-logic-based [11] and SPARQLent [21, 22]. The results show that the algorithms supported by PMRF behave globally well in comparison to iSeM-logic-based and SPARQLent frameworks.

In the future, we intend to enhance PMRF by (i) including other matching techniques; namely textual matching and Ontology distance calculation; (ii) adapt it to Ontology evolvement in a dynamic Web service environment; (iii) make the PMRF useable over the cloud technology; and (iv) use Artificial Intelligence techniques to reduce the cognitive effort required from the users/providers.

References

1. Arnold, K., O'Sullivan, B., Scheifler, R., Waldo, J., & Woolrath, A. (1999). *The Jini specification*. Reading, MA: Addison-Wesley.
2. Bellur, U., & Kulkarni, R. (2007). Improved matchmaking algorithm for semantic Web services based on bipartite graph matching. *IEEE International Conference on Web Services* (pp. 86–93). Salt Lake City, Utah, USA.
3. Ben Mokhtar, S., Kaul, A., Georgantas, N., & Issarny, V. (2006). Efficient semantic service discovery in pervasive computing environments. *ACM/IFIP/USENIX 2006 International Conference on Middleware* (pp. 240–259). Melbourne, Australia.
4. Chakhar, S. (2013). Parameterized attribute and service levels semantic matchmaking framework for service composition. *Fifth International Conference on Advances in Databases* (pp. 159–165), Knowledge, and Data Applications (DBKDA 2013) Spain: Seville.
5. Chakhar, S., Ishizaka, A., & Labib, A. (2014). QoS-aware parameterized semantic matchmaking framework for Web service composition. In: V. Monfort & K.H. Krempels (Eds.), *WEBIST 2014—Proceedings of the 10th International Conference on Web Information Systems and Technologies* (Vol. 1, pp. 50–61). Barcelona, Spain: SciTePress, 3–5 April 2014.
6. Doshi, P., Goodwin, R., Akkiraju, R., & Roeder, S. (2004). Parameterized semantic matchmaking for workflow composition. IBM Research Report RC23133, IBM Research Division
7. Forgy, C. (1982). Rete: A fast algorithm for the many patterns/many objects match problem. *Artificial Intelligence*, *19*(1), 17–37.
8. Gmati, F. E., Yacoubi-Ayadi, N., Chakhar, S. (2014). Parameterized algorithms for matching and ranking Web services. *Proceedings of the On the Move to Meaningful Internet Systems: OTM 2014 Conferences 2014* (Vol. 8841, pp. 784–791), Lecture Notes in Computer Science. Springer.
9. Gmati, F. E., Yacoubi-Ayadi, N., Bahri, A., Chakhar, S., Ishizaka, A. (2015). A tree-based algorithm for ranking Web services. In: V. Monfort, & K.H. Krempels (Eds.), *WEBIST 2015—Proceedings of the 11th International Conference on Web Information Systems and Technologies*. Lisbon, Portugal: SciTePress, 20–22 May 2015.
10. Guo, R., Le, J., & Xiao, X. (2005). Capability matching of Web services based on OWL-S. *Sixteenth International Workshop on Database and Expert Systems Applications* (pp. 653–657).
11. Klusch, M., & Kapahnke, P. (2012). The iSeM matchmaker: A flexible approach for adaptive hybrid semantic service selection. Web Semantics: Science, Services and Agents on the World Wide Web, vol. 15, pp. 1–14.
12. Klusch, M., Dudev, M., Misutka, J., Kapahnke, P., & Vasileski, M. (2010). SME^2 Version 2.2. User Manual. The German Research Center for Artificial Intelligence (DFKI), Germany.
13. Küster, U., & König-Ries, B. (2010). Measures for benchmarking semantic Web service matchmaking correctness. *Proceedings of the 7th International Conference on The Semantic Web: Research and Applications—Volume Part II* (pp. 45–59). ESWC'10, Berlin, Heidelberg: Springer.
14. Lee, C., Helal, A., Desai, N., Verma, V., & Arslan, B. (2003). Konark: A system and protocols for device independent, peer-to-peer discovery and delivery of mobile services. *IEEE Transactions on Systems, Man and Cybernetics, Part A: Systems and Humans*, *33*(6), 682–696.

15. Li, L., & Horrocks, I. (2003). A software framework for matchmaking based on semantic web technology. *Proceedings of the 12th International Conference on World Wide Web* (pp. 331–339). WWW '03, New York, NY, USA: ACM.
16. Ludwig, S. (2011). Memetic algorithm for Web service selection. *Proceedings of the 3rd Workshop on Biologically Inspired Algorithms for Distributed Systems* (pp. 1–8). BADS '11, New York, NY, USA: ACM.
17. Miller, B., & Pascoe, R. (2000). *Salutation service discovery in pervasive computing environments*. IBM Pervasive Computing: White paper.
18. Paolucci, M., Kawamura, T., Payne, T., & Sycara, K. (2002). Semantic matching of web services capabilities. *Proceedings of the First International Semantic Web Conference on The Semantic Web* (pp. 333–347). ISWC '02, London, UK: Springer.
19. Rodriguez-Mier, P., Pedrinaci, C., Lama, M., & Mucientes, M. (2015). An integrated semantic Web service discovery and composition framework. IEEE Transactions on Services Computing. Forthcoming.
20. Samper, Z. J., Llido, E. L., Soriano G. F., & Martinez D. J. (2015). Semantic Web service discovery system for road traffic information services. *Expert Systems with Applications*, *42*(8), 3833–3842.
21. Sbodio, M. (2012). SPARQLent: A SPARQL based intelligent agent performing service matchmaking. In B. Blake, L. Cabral, B. König-Ries, U. Küster & D. Martin (Eds.), *Semantic Web Services* (pp. 83–105). Berlin Heidelberg: Springer.
22. Sbodio, M., Martin, D., & Moulin, C. (2010). Discovering semantic Web services using SPARQL and intelligent agents. *Web Semantics: Science, Services and Agents on the World Wide Web*, *8*(4), 310–328.
23. Sharma, S., Lather, J., & Dave, M. (2015). Google based hybrid approach for discovering services. *IEEE International Conference on Semantic Computing (ICSC 2015)* (pp. 498–502). Anaheim, CA: IEEE.
24. Srujana, S., Raju, V., & Kiran, M. (2014). Semantic Web services discovery using logic based method. *Proceedings of the 3rd International Conference on Frontiers of Intelligent Computing: Theory and Applications (FICTA) 2014* (Vol. 1, pp. 623–629). Bhubaneswar, Odisa, India, 14–15 November 2014.
25. Sycara, K., Paolucci, M., van Velsen, M., & Giampapa, J. (2003). The retsina mas infrastructure. *Autonomous Agents and Multi-Agent Systems*, *7*(1–2), 29–48.
26. Syu, Y., Ma, S. P., Kuo, J. Y., & FanJiang, Y. Y.: A survey on automated service composition methods and related techniques. *The IEEE Ninth International Conference on Services Computing (SCC 2012)* (pp. 290–297). Hawaii, USA.
27. Wang, R., Chi, C. H., & Deng, J. (2009). A fast heuristic algorithm for the composite Web service selection. *Proceedings of the Joint International Conferences on Advances in Data and Web Management* (pp. 506–518). APWeb/WAIM '09, Berlin, Heidelberg: Springer.
28. Yu, T., & Lin, K. J. (2004). Service selection algorithms for Web services with end-to-end qos constraints. *Proceedings of the IEEE International Conference on e-Commerce Technology (CEC 2004)* (pp. 129–136).

Author Index

R. Lee (ed.), *Software Engineering Research, Management and Applications*, Studies in Computational Intelligence 654,
DOI 10.1007/978-3-319-33903-0